S0-AVN-523

Recent Developments in Scientific Optical Imaging

Recent Developments in Scientific Optical Imaging

Edited by

M. Bonner Denton, Robert E. Fields and Quentin S. Hanley
University of Arizona, Tucson, Arizona, USA

THE ROYAL
SOCIETY OF
CHEMISTRY
Information
Services

The Proceedings of the International Conference on Scientific Optical Imaging held in Georgetown, Grand Cayman on 29 November–2 December 1995.

Special Publication No. 194

ISBN 0-85404-786-7

A catalogue record for this book in available from the British Library

© The Royal Society of Chemistry 1996

All rights reserved.

Apart from any fair dealing for the purpose of research or private study, or criticism or review as permitted under the terms of the UK Copyright, Designs and Patents Act, 1988, this publication may not be reproduced, stored or transmitted, in any form or by any means, without the prior permission in writing of The Royal Society of Chemistry, or in the case of reprographic reproduction only in accordance with the terms of the licences issued by the Copyright Licensing Agency in the UK, or in accordance with the terms of the licences issued by the appropriate Reproduction Rights Organization outside the UK. Enquiries concerning reproduction outside the terms stated here should be sent to The Royal Society of Chemistry at the address printed on this page.

Published by The Royal Society of Chemistry,
Thomas Graham House, Science Park, Milton Road,
Cambridge CB4 4WF, UK

Printed by Bookcraft (Bath) Ltd

Preface

TK7871
.99
C45R42
1996
CHEM

Solid-state array detector technology is revolutionizing a host of disciplines. Astronomy, low-light level imaging, capture of low contrast images, and a wide range of spectroscopic applications are all benefiting from the use of these devices. Industrial interests have also noted the speed, sensitivity, dynamic range and wealth of information available from them. Applications are rapidly proliferating and spreading from the laboratory, to military, general research, routine plant process control, and even motion picture special effects.

This book represents a compilation of papers based on talks presented at the Third International Conference on Scientific Optical Imaging held November 29th through December 2nd, 1995 in Georgetown, Grand Cayman Island, British West Indies. The papers provide a glimpse of the exciting advances in scientific optical imaging, the current state of the art in available devices, commercial and academic research, development, and application.

The intended goal of the International Conference on Scientific Optical Imaging was to provide a means for developing communication links between widely diversified groups of scientists and engineers involved in the design and application of optical array sensors. The scientific program consisted of lecturers and posters presented by invited speakers. Plenary lectures given by world renowned experts in their field provided overviews of important areas in optical imaging with topics ranging from design considerations, device fabrication and integration, to applications and data reduction. Astronomers, spectroscopists, microscopists, chemists, device and camera system developers and manufacturers, instrumentation representatives, and optics experts all shared their knowledge.

The program and venue was designed to cut through communication barriers and establish this broad interdisciplinary forum. Lectures were held in the afternoon and evening which left mornings free for informal discussions in the leisure atmosphere of the tropical Grand Cayman Island. During the forums and free times, participants shared problems, solutions, and projected future trends in this rapidly progressing field. Manufacturers, developers and end users had the occasion to interact in mutually beneficial ways to shape the features of new devices. Users in one field had the unique opportunity to learn from users in another whose approach might be from an entirely different point of view.

The editors would like to thank all participants for attending, sharing their knowledge and experience and supplying these texts. We would especially like to thank Ms. Christina Jarvis for her help and extra effort in preparing this volume, Ms. Julia Schaper for her help in preparing the travel arrangements, and Ms. Jeanette Moscone for all of her on-the-spot assistance in seeing to the myriad requirements of a gathering of scientists attending a conference in paradise.

M. Bonner Denton   Quentin S. Hanley   Robert E. Fields

Contents

HIGH SPEED SCIENTIFIC CCDS

J. Janesick and A. Dingizian
Pixel•Vision
Advanced Devices and Systems Division
4952 Warner Ave., Suite 300,
Huntington Beach, California, 92649

G. Williams
Pixel•Vision
Commercial Products Division
15250 NW Greenbrier Pkwy
Beaverton, Oregon, 97006

M. Blouke
Scientific Imaging Technologies
P.O. Box 569
Beaverton, Oregon 97075

1 INTRODUCTION

Scientific CCD technology, now celebrating its twenty-fifth birthday, has matured to a remarkable degree. Through much of this period, CCD fabrication was beset by countless difficulties, including contamination, variable and incompletely understood processing steps, rudimentary design tools, deficient modeling of device performance, and inadequate testing and absolute characterization. As a consequence, a great deal of trial and error was involved. Technological progress took place despite these instabilities and unknowns but the price was high. As an example, a total of approximately 150 production lots were required to develop and produce suitable CCDs for Space Telescope and Galileo projects. In contrast, second generation CCDs currently in the Wide Field/Planetary Camera (WF/PC) aboard Space Telescope only required two production lots. The recent flight 1024 x 1024 Cassini Star Tracker CCDs were successfully produced on the first attempt. This maturation has also enabled the production of CCDs at prices that are more widely affordable. This in turn has spawned an explosion of new CCD applications ranging from Raman spectroscopy, phosphorescence spectroscopy, x-ray imaging, three-dimensional microscopy, and high volume separation techniques such as electrophoresis and thin-layer chromatography. Indicators of this maturity are numerous:

Charge Generation: The quantum efficiency of these devices is remarkably high (> 0.5) over unprecedented range of wavelengths, approximately 1-11,000 A. CCDs are applied throughout this spectrum which includes the NIR, visible, UV, EUV, and soft x-ray.[1-20] Special phosphor coatings directly applied to the CCD allow response into the hard x-ray range (i.e., > 10 keV).

New applications use CCDs to directly detect high energy particles (e.g., electrons and protons). For example, electron-bombarded CCDs (EBCCDs) offer low-light level performance superior to conventional image intensifier coupled CCD (ICCD) approaches.[21, 22] Backside devices are employed to eliminate ionizing radiation damage problems induced by particles such as electrons.[23-25]

Charge Collection: Device yield has improved where it is now commercially feasible to build wafer-scale arrays. For example, Scientific Imaging Technologies (SITe) is producing off-the-shelf wafer scale (4-inch) 2048 x 2048, 24-micron pixel rear illuminated CCDs.[26] They also have recently announced a 2048 x 4096, 15- micron pixel thinned CCD that is buttable on three sides (a new sensor discussed in these proceedings). A 7k x 9k, 12-micron pixel CCD built by Philips is currently under test. A single chip occupies a six-inch silicon wafer (active region = 10.8 cm x 6.4 cm). Recent production yields have progressed to the point where CCDs having 10^8 pixels are being planned.

Design, process and clocking improvements have increased well capacity almost a factor of ten compared to the first scientific CCDs fabricated (e.g., WF/PC I compared to WF/PC II). Such performance has yielded CCDs that exhibit a dynamic range greater than 10^6, with near ideal linear response.

Charge Transfer: Charge transfer efficiency for the CCD is nearly perfect. Silicon quality today allows outstanding CTE performance to levels where it is difficult to characterize and measure. CTE's better than 0.999999 per pixel transfer are consistently measured for small charge packets (< 2000 e⁻) without fat-zero.[27, 28]

Charge Detection: Read noise below one electron rms has been finally achieved for slow-scan operation.[29, 30] Such noise performance has opened up many new applications where the CCD now competes with PMT detectors. Sub-electron noise performance allows x-ray imaging spectrometers to exhibit "Fano-noise energy resolution" over the entire soft x-ray range from 100 to 10 keV.[31, 32]

CCD sensors that achieve the performance above are referred to as "scientific slow-scan CCDs". For optimum performance slow-scan CCDs usually require a significant amount of time for charge integration and readout (times measured in seconds, minutes and even hours per frame time). These long staring times are used for on-chip integration of weak signals (as low as a few electrons per hour in the case of Space Telescope).

As recently as five years ago, more than 90% of high performance CCDs were used in slow-scan applications. However, the situation has changed dramatically and we now see an explosion of speed-sensitive applications. Many of these are biological imaging tasks in which higher frame rates are required to follow dynamic events in real time. This emphasis on speed has been prompted by advancements made in computer technology. Older computers just a few years back were unable to handle data-acquisition rates faster than slow-scan rates (50kHz - 1 MHz). Now faster PCs and workstations can cope with 16-bit data at rates in excess of 10 MHz. For example, Pixel▪Vision has developed a CCD camera system that is PCI based and can acquire data from multiple output nodes at rates exceeding 40 MHz.

The new class of CCDs being developed for high-speed applications, the subject of this paper, is referred to as "high-speed scientific CCDs (HSS CCDs)".

2 HIGH SPEED PERFORMANCE

CCD speed is defined by the pixel readout rate and frame rate. The pixel readout rate is the total number of individual readout events that occur per second. The frame rate is the number of complete frames, or images, read out per second. Frame rate is related to the readout rate, the number of pixels, and the extent of any desired binning (on-chip combining of pixel data). It also includes various contributions to time overhead such as register shift time and shutter open/close times (e.g., the four 800 x 800 Space Telescope CCDs are read out at 50 kpixels/sec and require several minutes for a complete picture).

For area array CCDs clock speed is specified for both the vertical and horizontal registers. Vertical clock speed is specified by the amount of time required to transfer charge (at full well) one full line (given as micro-sec/line). This specification is important for frame transfer devices where signal charge must be transferred as fast as possible from the image to the storage regions of the sensor (for smear reasons). In the horizontal direction we specify the time required to move charge by one column (micro-sec/column). This parameter typically represents the highest clock frequency applied to the CCD (often given as pixels/sec). Scientific CCDs have typically been operated at a 50 kpixel/sec data rate. In this paper we will define a High Speed Scientific CCD as one that has an effective data rate of $\geq$ 1Megapixel/sec.

High-speed readout results in degraded performance for the CCD compared to slow-scan operation. In fact, virtually every CCD performance parameter degrades when the clocking rate to the sensor is increased. To maintain optimum performance at higher speeds, slow-scan CCD technology requires modification in terms of how the sensor is designed and processed. We will briefly review some high-speed considerations currently under development at CCD manufacturers in areas of charge generation, collection, transfer and detection.

2.1 Charge Generation

High-speed CCDs require small pixels (< 12-microns). Small pixels exhibit low-clock drive capacitance and the inherent ability to transfer charge quickly from phase to phase. This virtue to transfer charge rapidly is due to the presence of higher fringing fields in the channel and a shorter distance for the charge packet to travel, a subject discussed below.[33] However, as the pixel size is reduced the sensitivity of the CCD degrades because "overhead structures" occupy a larger fraction of the pixel. The gate-to-gate poly overlap between phases is a good example of an overhead structure. Incident photons are absorbed in these overlap regions, lowering QE performance. The channel stop is another example of an overhead structure where photo electrons recombine, resulting in lower sensitivity.

For a given CCD process, overhead geometry remains fixed as the size of the pixel is reduced. For very small pixels the QE loss can be significant. For example, a 7.5-micron pixel exhibits a peak QE of only 30% when standard three-phase processing is employed. In comparison, a 24-micron pixel is nearly a factor of two greater using the same process.[27]

Gate-to-gate overlaps and channel stop losses for high-speed/small pixel devices can be circumvented by using backside illuminated CCD technology. In addition, because the back surface of a thinned CCD is of uniform composition, antireflection (AR) coatings can be employed (AR coatings cannot be used on frontside illuminated sensors). This is an important advantage given the high refraction index of silicon (> 3.5 at visible and near-IR wavelengths). This high index would otherwise produce large reflection losses of more than 30%, resulting in lower QE. With even the simplest single layer AR coating the QE can be pushed as high as 90% at the device's peak response wavelength (which can be tailored by adjusting the thickness of the coating). This results in larger signals, and hence, improved signal-to-noise performance. In low-light applications, this feature can be used to generate superior-quality data and/or to increase frame rate.

An increasing number of applications rely on imaging of auto-fluorescence from organic and aromatic molecules requiring UV sensitivity.[34, 35] Some CCD manufacturers extend the sensitivity for front-illuminated CCDs into the UV region using a UV-sensitive phosphor to "shift" the wavelength of incident UV light (e.g. Space Telescope CCDs), but this approach produces limited quantitative utility and low overall efficiency (especially for small pixel devices where overhead structures again take their toll). Back-illuminated devices, however, can support direct UV detection with high quantum efficiency. In fact, special UV-enhanced AR coatings developed at SITe yield back-illuminated devices with a QE of approximately 50% throughout the UV, whereas the QE for conventional front-illuminated CCDs drops to zero at 380 nm.[9]

2.2 Charge Collection

Many years ago, when CCDs were being built for Space Telescope, it appeared that arrays greater than 800 x 800 pixels would be difficult to fabricate. It was doubtful that a 1024 x 1024 sensor could be successfully built without exhibiting some cosmetic features. Today, the 1024 x 1024 sensor is a standard CCD format, proving earlier predictions completely wrong. After the 1024 x 1024 CCD was fabricated with success, larger arrays were immediately put on the drawing board, leading to arrays we see today. For example, hundreds of 2048 x 2048 CCDs have been fabricated and sold. At least two CCD manufacturers are now fabricating 4096 x 4096 formats, used primarily for mammogram imaging applications, but, of course, also popular with astronomers. Currently Loral is making a 9k x 9k CCD on a 5-inch silicon wafer (discussed in these proceedings). Philips has just finished fabricating a 7k x 9k, 12-micron pixel device built on a six-inch wafer. It is now anyone's guess how large the CCD will grow, considering that silicon semiconductor manufacturers are just beginning to use 12-inch silicon wafers.

The super big arrays described above are forcing slow-scan users into the high-speed world whether they know it or not. For example, the new 7k x 9k Philips CCD requires 21 minutes of read time when read out at 50k pixels/sec through a single amplifier. Fortunately, the sensor incorporates four output amplifiers which can be read out in parallel, thus reducing read time to 315 seconds (5.25 minutes). In addition, the readout rate can be increased to 250k pixels/sec and still achieve the same 50 kpixels/sec noise floor by using new high-speed ADCs. Hence, total read time for the CCD can be 63 seconds, considerably shorter than 21 minutes. At this rate the CCD will be effectively reading at 1000 kpixels/sec using "parallel processing" techniques (at this speed, the CCD is considered a HSS CCD).

Well capacity rapidly decreases for pixel dimensions less than approximately 15-microns. Therefore, careful consideration must be given to the design and processing aspects of a HSS CCD pixel. For example, overhead functions (e.g., channel stops) need to be designed and fabricated as small as possible. Also in processing small pixel devices, the depth and doping concentration of the n-buried channel must be carefully chosen to maximize well capacity. The quantity of charge that can be collected in a pixel is directly proportional to the doping level of the n-layer. However, there is a practical limit to the channel doping concentration that can be employed. For example, increasing the doping level requires greater clock swings in controlling charge transfer. This in turn increases the stress across the gate insulator, potentially making the device less reliable. However, it currently appears that the main doping limitation is not clock swings but is instead associated with high internal electric fields that are generated with increased doping concentration. Excessive doping can result in a weak avalanche condition in the middle of the signal channel, which leads to dark spike generation (pixels that exhibit very high dark current generation).[27, 36]

Channel doping also influences performance characteristics of the on-chip amplifier, since it is typically doped at the same time.[37] The performance of this transistor degrades with increasing channel doping, resulting in higher read noise, lower transconductance and nonlinearity problems. To circumvent the difficulty, some CCD manufacturers now dope the amplifier and array regions of the sensor independently, via different reticles. In this manner, full well and noise characteristics can be optimized separately. The difference in buried channel doping between the array and amplifier for improved full well performance has recently been shown to be significant (1.9×10^{16} cm^{-3} and 1.3×10^{16} cm^{-3}, respectively).[27] Although not fully optimized, work in this area continues to show enhanced full well performance and better amplifier characteristics. These results are critically important for small pixel devices if high full well is to be maintained.

MTF performance is usually worse for small pixel CCDs, because photoelectrons can wander from a small target pixel more readily than from a larger one. Recall that MTF is primarily limited by the amount of neutral (i.e., field free) material below the frontside depletion edge. For front-illuminated CCDs, MTF loss occurs for wavelengths greater than 700 nm, where photons can penetrate below the depletion region into substrate material below. Hence, epi-thickness and resistivity are critical parameters for optimum MTF performance. Ideally, the depletion edge should extend very close to the epitaxial interface to minimize charge diffusion effects.

For back-illuminated devices the situation is more critical. For this technology the CCD should be thinned to the depletion edge. Any field free material left after thinning can significantly influence MTF performance, especially for small pixel devices. Wavelengths that exhibit short absorption depths, such as UV, EUV and soft x-ray, will suffer the greatest penalty because charge is generated at the surface of the device, leading to charge diffusion effects and MTF degradation.

MPP operation can exhibit degraded MTF performance compared to conventional CCDs. When all phases to a CCD are inverted, as is the case for MPP operation, the extension of the depletion region is minimized maximizing the extent of the field free region. For reasons discussed above this condition leads to reduced MTF. In comparison, the depletion edge extends deeper into the silicon when phases are biased high, the condition that occurs for partial or noninverted operation. Again, a thinned MPP CCD will suffer the most from this effect. Also, MPP CCDs always exhibit lower well capacity than conventional CCDs. For large pixel devices this problem is typically not a concern. However, as mentioned above, full well is a premium for small pixel sensors. As we shall discuss in the next section, MPP CCDs also have difficulty transferring charge quickly.

Therefore, three problems are associated with high speed MPP CCDs. MPP operation is not critical for these applications because low dark current generation is not normally an issue. Hence, MPP technology will not be employed in high speed designs.

2.3 Charge Transfer

Charge transfer efficiency (CTE) problems can potentially develop for high-speed thinned CCDs. This is because the substrate, which normally acts as ground plane for front-illuminated CCDs, is lost in the thinning process. Hence, the ground impedance between the front and back surfaces of the CCD significantly increases resulting in speed problems not encountered for nonthinned operation. The CTE effect is more pronounced for large array devices (> 1024 x 1024 CCD working at 30 frames/sec). The current solution under investigation for this problem is to deposit a thin layer of optically transparent metal on the back surface to reestablish the ground plane for high-speed operation, i.e., flash gate.

Well capacity decreases as the clock frequency to the CCD is increased.[27] Curiously, the effect observed is a CTE problem related to "fringing fields". Fringing fields develop when there is a potential difference between phases. That is, potential on one gate will induce a fringing field under the edges of neighboring phases and in turn "pull" charge out of the phase. Without fringing fields, charge only moves from phase to phase by thermal diffusion aided by self-repulsion, both of which are relatively weak transfer processes.[33] Fringing fields can extend several microns into neighboring phases when charge is absent, exhibiting maximum strength. However, as charge collects the potential difference between phases is lowered, resulting in smaller fringing fields. For this reason, it is much easier for a CCD to transfer smaller charge packets than large ones. At full well, fringing fields essentially disappear and charge transfer can only rely on diffusion and self-repulsion forces. Under full well conditions, if the clocking rate is too fast, transfer will be incomplete and charge will bloom backwards. For example, a three-phase, 12-micron pixel requires about 1 micro-sec/line to transfer a full well signal (100,000 e⁻) without exhibiting the blooming effect.[27] In comparison a 1600 e- charge packet can be transferred in less than 50 ns for the same CCD. The CCD can be made to run faster by shortening the length of a phase to reduce the diffusion time and increasing the fringing field strength between phases. For example, a 7.5-micron three-phase pixel can transfer a full well signal in less than 200 ns.

It should also be mentioned that large pixels can be designed to operate high-speed. We refer to such pixels as "super pixels". For example, six-phases, each 4-microns long, can be employed to make a 24-micron pixel (using 3 levels of poly). A conventional three-phase, 24-micron pixel (i.e., 8-micron gates) would require significantly more time to transfer charge because transfer time varies exponentially with gate length.[27]

Other CTE difficulties become apparent when the CCD is clocked faster. For example, "traps", which for the most part have been cured for the slow-scan CCD, become critical once again for the HSS CCD.[38] Electron traps come in many flavors and are manifestations of incorrect CCD design or process anomalies or poor silicon quality. Thermal diffusion and transfer time are the main mechanisms that allow signal charge to escape from traps. More transfer time is required for deeper traps to keep signal charge in the target pixel. However, as the CCD is clocked faster, even shallow traps, not critical to slow-scan operation, begin to limit CTE performance, especially those situated in the horizontal register where speed is greatest.[39]

The main solution to the high-speed trap problem is to carefully screen, via x-ray stimulation. each device at the pixel rate at which the device will be used. The user with low-light level traps can alleviate the problem by maximizing clock overlap periods, thereby allowing charge to escape from trapping sites. Also, "fat-zero" can be introduced to fill traps to a level where shot noise becomes too great.

MPP CCDs clocked fast can exhibit CTE problems. The difficulty has been traced to the MPP implant itself. During high temperature processing the MPP implant encroaches under adjacent phases (typically phases 1 and 2 for a three phase CCD). This effect decreases the potential at the edges of these phases, generating a small but influential potential barrier to electrons. When phase 3 is empty, these barriers are overwhelmed by fringing fields and charge transfer is well behaved. However, as signal electrons are collected, the fringing fields collapse and the barriers become apparent, making it more difficult to transfer charge. Therefore, the full effect of this problem becomes present at full well conditions where fringing fields are smallest. Charge can only escape over the barriers by thermal diffusion. Since the potentials associated with the barriers are probably many times greater than kT/q, significantly more transfer time is required compared to CCDs without the MPP implant.

2.4 Charge Detection

Currently read noise for the best CCDs fabricated increases from approximately 2 e⁻ rms at 100 kpixels/sec to approximately 20 e⁻ rms at 10 Mpixels/sec (i.e., a factor of 100 in speed results in a factor of 10 in noise). Unfortunately, significant improvements in output amplifier noise are not likely, based on theoretical arguments. However, some optimization still remains to be performed. For example, it appears that an optimum amplifier geometry exists depending on pixel rate employed. As the pixel rate increases the processing time given to each pixel becomes shorter (we assume in this discussion that "correlated double sampling" is employed). Reducing the sample time lowers 1/f noise generated by the output amplifier because low-frequency noise components become correlated. Reducing 1/f noise through signal processing allows one to reduce the size of the on-chip amplifier. This in turn increases the sensor's sensitivity (V/e⁻) because of reduced gate capacitance associated with the MOSFET amplifier. Noise, i.e., electron rms, is reduced proportionally. However, as the amplifier becomes smaller, its transconductance (V/A) decreases, resulting in a greater white noise floor. This results in higher noise and sets the limit on how small the device can be. Therefore for a given sample time, the point where white and 1/f noise increase more than a sensitivity increase defines the optimum amplifier size.[40, 41]

Curiously we still don't know the location from which 1/f noise is generated. In that scientific CCD amplifiers employ buried channel MOSFETs, surface state noise (or 1/f) should not be apparent, according to theory. However, the noise seen is substantial, compared to, say, JFETs. Somewhere along the channel of the MOSFET drain, current must interact with the surface to generate 1/f noise. One possibility is that charge "jumps" from the channel into surface states thermally, possibly near the source region. We know from surface full well measurements that electrons can obtain sufficient energy to overcome a barrier height of several tenths of a volt (i.e., kT). Reducing operating temperature does in fact reduce 1/f noise, making this theory attractive. However, reducing temperature also increases the emission time constants of the surface states. This reduces the frequency components, effectively lowering 1/f noise, because correlation is increased. Measurements performed on SITe CCDs exhibit a much lower 1/f noise floor compared to other CCD manufacturers. SITe's process involves many high temperature steps which drive-in the channel, possibly reducing surface interaction. On the other hand, the SITe process could also be passivating the surface states more effectively than other CCD groups; surface dark current is very low for the SITe process. As can be seen, there are many questions about 1/f noise and where it truly originates. Further work in this area needs to be performed.

There are other techniques available that effectively reduce noise for high-speed applications. One direct approach relies on parallel processing where multiple readout amplifiers are employed. Using this technique the noise is reduced (or speed increased) by the square-root of the number of amplifiers employed. This approach has been successfully implemented for high-speed CCDs made today that exhibit ultra-low noise performance. For example, as we will discuss in the next section, some HSS CCDs are being read out at 2-3 thousand frames/sec and at the same time achieve 3 e- noise performance using parallel processing techniques.

3 SANDBOX CCDS

A "Sandbox Lot" run is an effort to capitalize upon current CCD technology at a reduced cost in CCD development and production.[27] In doing so, the approach enables development of custom CCDs the cost of which would have been prohibitive separately, and also facilitates research of important CCD issues, such as high-speed optimization, at greatly reduced cost. These savings are realized by combining several devices on a single development lot, thus dividing the production costs among several customers.

A recent Sandbox run fabricated at Reticon has produced several custom CCDs intended for high-speed use. Some CCDs currently under test are as follows:

Cinema CCD: The Cinema CCD is a first generation 4096 x 4096, 3-phase, 9-micron pixel, frame transfer, color CCD. This HSS CCD is intended to replace photographic film cameras used in making moving pictures. The prototype CCD offers several new design challenges. For example, the CCD will be read out at 30 frames/sec at an effective pixel rate of a quarter billion pixels per second. To achieve such rates the CCD is divided into 32 sections for parallel readout. The array is vertically split into four 1024 (V) x 4096 (H) sections, the middle two sections forming two image and and the outer regions, two storage sections. The split image is transferred towards the top and bottom of the array (such a layout is referred to as "split-frame-transfer". From each storage region there are 16 individual horizontal registers each that report to a 3-stage on-chip amplifier. Each channel reads at approximately 8 M pixels/sec. The resulting signals are encoded to 12-bits to be later compressed.

Mach II CCD: The Mach II CCD is a second generation 1024 x 1024, 6-phase, 36(V) x 18(H)-micron super pixel, interlaced-frame transfer CCD. The sensor will be used to take two sequential 512 (V) x 1024 (H) images separated in time by less than 1-micro-sec. Phases 4, 5, and 6 are masked off with a light shield serving as a storage region for each line. The first image is collected in phases 1-3 and then quickly transferred (<1-micro-sec) into storage phases 4-6. A second image is then taken. The two interlaced images are then read out slow-scan (50 kpixels/sec) to achieve 4 e-noise performance. Six-phase design is employed for high QE performance, low optical cross-talk between the two images and high-speed operation. The CCD also incorporates a high-speed readout channel for fast-scan applications where low-noise performance is not a requirement.

Adapt II CCD: The Adapt II CCD is a second generation 128(V) x 64 (H), 6-phase, 36-micron pixel, frame store, CCD. The CCD is used for wavefront sensing in Adaptive Optics camera systems. The device will scan between 2000 - 3000 frames/sec at a read noise < 3 e- rms. Sixty-four amplifiers are provided, one amplifier per column. The amplifiers are Skipper type to allow for multiple sampling and low-noise.

Pluto Flyby CCD: The Pluto Flyby CCD is a first generation 2048 (V) x 1024 (H), 3-phase, 9-micron pixel, frame store CCD. The sensor will potentially be used in a NASA planetary mission to planet Pluto. One simple floating diffusion amplifier is used to read the array. A second generation Pluto device is described below for high-speed operation.

There are also two Sandbox lots currently being built at Loral Fairchild. These CCDs have the following characteristics:

007: The 007 device is a "standard video format" thinned CCD that will run at 30 frames/sec (intended for night vision applications). The imaging region is 650 pixels in the horizontal direction and 489 in the vertical with anti-blooming. The memory section is 650 x 499 without anti-blooming. The pixels in the imaging section are 13.75 x 13.75-microns. In the memory the pixels are 9 x 13.75-microns. The vertical registers are based on three phases per pixel while the horizontal register uses two-phase technology for high-speed readout (> 10 Mpixels/sec). The single on-chip amplifier is a buried channel two-stage MOSFET with on-chip load.

Adapt III: Adapt III is similar to the Adapt II sensor described above with the exception that floating diffusion amplifiers are used.

Confocal I: This CCD will be used in conjunction with laser confocal microscopes (CFM) which is a new research tool in biology and materials science.[34,42,43] For biological applications, it is usually employed to detect the location of fluorescent marker molecules. The instrument is a scanning-laser microscope with the CCD mounted behind a spatial filter aperture which is placed conjugate to the "reflected" image of the spot. This creates an image of an optical section, such as a living cell.

The prototype CCD is the same as Adapt III except the sensor incorporates a traditional, 64-stage horizontal register at the top of the sensor that is coupled to a single 36 x 36-micron diode region used to detect the confocal spot of light. In operation, charge is sampled and shifted out of the diode at approximately 1- 2 MHz rate, filling the horizontal register with discrete samples. When full, the 64 charge packets from the horizontal register are shifted through the 128 stage vertical registers to the 64 amplifiers. Parallel processing maintains low noise for the device (< 3 e-rms). The resulting detector is more compact, efficient and reliable than the photo multiplier tube (PMT) it replaces .

BIG CIT: A 4096 x 4096, 15-micron pixel CCD to be applied in astronomical applications. Four amplifiers, one in each corner, will read the CCD at approximately 1 M pixel/sec. The device will be thinned and backside illuminated.

Two Sandbox runs are planned at SITe this year to produce four custom devices. All sensors will be thinned, accumulated and AR coated using SITe's standard thinning process. These devices are as follows:

Confocal II: Similar to Confocal I except that only 16 readout channels are employed. Also, the horizontal register reports directly to the amplifiers without vertical registers. In addition, a 3 x 3 frame transfer array replaces the Confocal diode. The subarray will be read out at 1-2 Mframes/sec. Such an arrangement can be used as a quadrant detector to measure optical misalignments of the microscope. Also by looking at the pixel with the most signal one can track the center of the confocal signal spot in order to compensate for the small lateral beam deflections caused by optical homogenities in the specimen. Alternatively, one could use the 3 x 3 region and change its sampling pattern to effectively vary the pinhole size electronically.[43]

Pluto II: Similar to Pluto I except four high-speed amplifiers (one in each corner) will be incorporated including split-frame-transfer. The CCD will be read out at approximately 30 frames/sec.

Adapt IV: Similar to Adapt III except it will be a 160(V) x 80(H) split-frame-transfer CCD. There are a total of 80 amplifiers, two amplifiers per column.

Fast: A 160(V) x 80(H) ultra-fast-split-frame-transfer CCD based on 18-micron pixels. Four high-speed amplifiers are located in each corner.

ACKNOWLEDGMENTS

We thank Jeff Pinter of Loral Fairchild and Rusty Winzenread of EG&G Reticon for many rewarding technical discussions on CCD technology. We also thank Tom Elliott of the Jet Propulsion Laboratory for help in acquiring test data that supports many theoretical discussions in this paper.

References

1. M. Blouke, J. Janesick, T. Elliott, S. Collins and J. Freeman, *Optical Engineering*, 1987, **8**.

2. J. Janesick and T. Elliott, "History and advancement of large area array scientific CCDs," Astronomical Society Pacific Conference Series '91, Sept. 1991, Tucson, AZ.

3. J. Janesick, S. Collins, T. Elliott and H. Marsh, "The future scientific CCD", in State of the Art Imaging Arrays and Their Applications, Proc. SPIE, Aug. 1984, Vol. 501.

4. J. Janesick, T. Elliott, T. Daud, J. McCarthy and M. Blouke, Backside charging of the CCD", in Solid State Imaging Array Conference, Proc. SPIE, Aug. 1985, Vol. 570-78, p. 46.

5. J. Janesick, T. Elliott, T Daud and D. Campbell, "The CCD flash gate", in Instrumentation in Astronomy VI, Solid State Imaging Arrays for Astronomy, March 1986, Tucson, AZ.

6. J. Janesick, T. Elliott, G. Fraschetti, S. Collins, M. Blouke and B. Corrie, "Charge coupled device pinning technologies" in "Optical Sensors and Electronic Photography, Proc. SPIE, Vol.1071-15, p. 153.

7 M. Blouke, J. Janesick, J. Hall, M. Cowens and P. May, *Optical Engineering*, 1983, **22**, 607.

8. J. Janesick, S. T. Elliott, T. Daud and D. Campbell, *Optical Engineering*, 1987, **26**, 852.

9. M. P. Lesser, *Optical Engineering*, 1987, **26**.

10. M. Blouke and M. D. Nelson, "Stable ultraviolet antireflection coatings for charge-coupled devices" in Charge-coupled Devices and Solid-State Optical Sensors III, ed. M. Blouke, Proc. SPIE, 1993, Vol. 1900.

11. M. Cowens, M.Blouke, T. Fairchild and J. Westphal, *Applied Optics*, 1980, **19**, 3727.

12. R. A. Stern, R. C. Catura, R. Kimble, A. F. Davidsen, M. Winzenread, M. Blouke, R. Hayes, D. M. Walton and J. L. Culhane, *Optical Engineering*, 1987, **26**, 875.

13. K. Marsh, C. Joshi and J. Janesick, "CCD spectroscopy of plasmas", IEEE International Conference on Plasma Science, May 1984, St. Louis, Missouri.

14. J. Janesick, T. Elliott, H. Marsh, S. Collins, M. Blouke, and J. McCarthy, *Rev. of Sci. Instrum.*, 1984, **56**, 796.

15. J. Janesick, T. Elliott, J. McCarthy, S. Collins and M. Blouke, *IEEE Transactions Nuclear Science Symposium*, 1985, **NS-32**, No. 1, 409.

16. R. Stern, K. Liewer and J. Janesick, *Review of Scientific Instrumentation*, 1983, **54**, 198.

17. M. Blouke and J. Janesick, *Optical and Photonics News*, April 1995, **6**, 16.

18. K. Marsh, J. Janesick, C. Joshi and S. Collins, *Review of Scientific Instruments*, 1985, **56**, 837.

19. J. Janesick, D. Campbell, S. Collins, T. Daud, T. Elliott and G. Garmire, "CCD advances for x-ray scientific measurements in 1985", in X-ray Instrumentation in Astronomy, Proc. SPIE, 1985, Vol. 597.

20. T. Daud, J. Janesick, T. Elliott and K. Evans *Optical Engineering*, 1987, **26**, 686.

21. G.M. Williams and M. M. Blouke, *Laser Focus World*, 1995, **31**.

22. G. Williams, A. Reinheimer, V. Aebi, and K. Costello, "Electron-bombarded backside-illuminated CCD sensors for low-light-level imaging applications", in Charge-coupled Devices and Solid-state Optical Sensors IV, ed. M. Blouke, Proc. SPIE, 1995, Vol. 2415.

23. J. Janesick, F. Pool and T. Elliott, "Radiation damage in scientific CCDs", IEEE Nuclear Science Symposium, Nov. 1988, Orlando, FL.

24. L. Acton, M. Morrison, J. Janesick and T. Elliott, "Radiation concerns for the Solar-A Soft X-ray Telescope", in CCD and Solid State Optical Sensors II, Proc. SPIE, 1991, Vol.1447-11.

25. J. Janesick, G. Soli, T. Elliott and S. Collins, "The effects of proton damage on charge coupled devices", in Charge Coupled Devices and Solid State Optical Sensors II, Proc. SPIE, 1981, Vol. 1447, p. 87.

26. M. Blouke, B.Corrie, D. Heidtmann, F. Yang, M. Winzenread, M. Lust and J. Janesick, *Optical Engineering*, 1987, **26**, 837.

27. J. Janesick, T. Elliott, R. Winzenread, J. Pinter and R. Dyck, "Sandbox CCDs", in Charge-coupled Devices and Solid-state Optical Sensors IV, ed. M. Blouke, Proc. SPIE, 1995, Vol. 2415, p. 2.

28. J. Janesick, R. Bredthauer, J. Pinter and L. Robinson, "Notch and Large Area CCD Imagers", in <u>CCD and Solid-state Optical Sensors II</u>, Proc. SPIE,1991, Vol. 1447.

29. J. Janesick, T. Elliott, A. Dingizian, R. Bredthauer, C. Chandler, J. Westphal, and J. Gunn, "New advances in charge-coupled device technology - sub-electron noise and 4096x4096 pixel CCDs", in <u>Charge-coupled Devices and Solid-state Optical Sensors</u>, ed. M. Blouke, Proc. SPIE, 1990, Vol. 1242.

30. C. Chandler, J. Gunn, J. Janesick, J. Westphal and R. Bredthauer, "Sub-electron noise charge coupled devices", in <u>Symposium on Electronic Imaging</u>, Proc. SPIE, 1990, Vol. 1242.

31. R. Kraft, J. Janesick, D. Burrows, G. Garmire , J. Nousek, M. Skinner, and D. Lumb, "Soft x-ray spectroscopy using charge coupled devices with thin poly gates and floating gate output amplifiers," Proc. SPIE, 1994, Vol. 2278.

32. J. Janesick, T. Elliott, R. Bredthauer, C. Chandler and B. Burke, "Fano-noise limited CCDs", in <u>Optical and Optoelectronic Applied Science and Engineering Symposium, X-ray Instrumentation in Astronomy</u>, Proc. SPIE, San Diego CA, 1988.

33. J. Carnes, W. Kosonocky, and E. Ramberg, *IEEE Jour. of Solid-State Circuits*, 1971, **SC-6**, 322.

34. J. Pawley, 'Handbook of Biological Confocal Microscopy', Plenum, NY, 1995.

35. D. Agard, Y. Hiroaka, P. Shaw and J. Sedat, *Molecular Cell Biology*, 1989, **30**, 353.

36. For a discussion of CCD fabrication see, for example, 'Charge-Coupled Devices and Systems', eds. M. J. Howes and D. V. Morgan, Wiley and Sons, New York, 1979.

37. J. Pinter, J. Janesick, J. Bishop, T. Elliott, "2-D Modeling of CCDs Optimum Design and Operation for Maximum Charge Handling Capability," Proc. SPIE, Dana Point, CA, 1995.

38. M. Blouke, D. Heidtmann, F. Yang and J. Janesick, "Traps and deferred charge in CCDs", in <u>Instrumentation for Ground-based Optical Astronomy: Present and Future</u>, ed. L. B. Robinson, Springer Verlag, New York, 1988, p.462.

39. J. Flores, "CTE model for estimating image smear", in <u>Photo electronic Detectors, Cameras and Systems</u>, eds. C. Johnson and E. Fenyves, Proc. SPIE, 1995, Vol. 2551, p. 308.

40. H. Kim and D. Heidtmann, "Characteristics of 1/f noise of the buried-channel charge-coupled device (CCD)", in <u>Optical Sensors and Electronic Photography</u>, eds. M. Blouke and D. Pophal, Proc. SPIE, 1989, Vol. 1071, p. 66.

41. P. Centen, *IEEE Trans. on Electron Devices*, 1991, **38**, 1206.

42. J. Pawley, " Detectors for 3D microscopy", in <u>Confocal Microscopy</u>, eds. Stevens, et al., Academic Press, New York, 1994, p. 48.

43. J. Pawley, J. Janesick, and M. Blouke, "The CCDiode: An optimal detector for laser confocal microscopes", in <u>Three dimensional microscopy: Image acquisition and processing III</u>, eds. C. J. Cogswell, G. S. Kino, and T. Wilson, Proc. SPIE, 1996, Vol. 2655.

A NEW LARGE FORMAT, VERY LOW NOISE SCIENTIFIC CCD

P.J. Pool

EEV Ltd.
Waterhouse Lane
Chelmsford   CM1 2QU

D.J.Burt.

GEC-Marconi Materials Technology
Hirst Division, Elstree Way
Borehamwood WD6 1EX

P.R. Jorden and A.P. Oates

Royal Greenwich Observatory
Madingley Rd.
Cambridge CB3 OEZ

1 INTRODUCTION

Astronomical telescopes and associated instruments have focal plane sizes up to 100mm diameter so the efficient use of this area requires a large sensor or sensor array with high quantum efficiency and low read noise. To meet this requirement, in a collaboration between the Royal Greenwich Observatory (RGO), Anglo-Australian Observatory (AAO), national Galileo Telescope (TNG) and EEV, a large area, back illuminated, 3-edge buttable CCD has been designed with very low noise output circuits.

The primary design aim was to meet the requirements of the users, but an effort was also made to put in place enabling structures for experimental device modifications to allow for future performance enhancement.

2 REQUIREMENT

The definition of the device format and performance requirement was agreed between the users (RGO, AAO and TNG) and the manufacturer (EEV). To aid the process of design trade-off, several parameters have both specification limits and targets (the target figures being shown in parenthesis in the following text):

Image format	2048(H) x 4096(V)
Register format	Split, with gated dump drain
Pixel size	13.5μm x 13.5μm
Inactive edge space	
(Active area to Si edge)	
-top	100μm [50μm]
-sides	500μm [250μm]
Pixel full well	150k electrons [200k]
Register full well	x 2 pixel [x4]

Output node full well
(summing well and node)   x 4 pixel [x9]

Max. parallel transfer rate   _ 20kHz

Max. serial transfer rate   _ 1MHz

Quantum efficiency   Wavelength nm.   QE%
   350   15   [50]
   400   50   [70]
   650   70   [85]
   900   25   [40]

Output circuit noise   <4 electrons rms @ 50kHz
[<2 electrons @ 50kHz]
[<5 electrons @ 1 MHz]

3 DESIGN

3.1 Format

Figure 1 shows a schematic diagram of the CCD42 chip. This is of conventional "stitched"[1] format with a "middle" section comprising 2048(H) x 512(V) square pixels on 13.5μm pitch and an "ends" section comprising the top and bottom terminations for the array, including 4 extra lines of pixels, the read-out components and the wire bonding pads. The primary variant is to have eight middle sections (hence the type number CCD42-80) giving a total of 2048(H) x (4100)V pixels and an image area of approximately 27.6mm x 55.3mm. Other variants with different numbers of middle sections can of course be produced. To facilitate thinning for back face imaging, gate protection circuits are designed close to the device and the bond pads are spaced off.

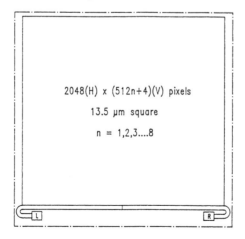

Figure 1 *Schematic diagram of CD42 chip*

3.2 Image Section

From the design of the image section it is expected that the peak signal level will be 170k electrons typical, 150k electrons minimum. The dark current should be about 1nA/cm² at 20°C, reducing with the usual exponential dependence on temperature. An Advanced Inverted Mode variant would support a peak signal level of ≈ 100k electrons.

3.3 Read-out Register

Separate charge detection circuits are incorporated at each end of the read-out register, which is split such that a line of charges can be transferred to either output, or split between the two. The register is provided with a drain and control gate along the outer edge of the channel for charge dump purposes. The application of about +10V (not critical) to this control gate causes charge to flow from the register directly to the drain, allowing fast clearing of charge from all or part of the image section. Measurements of a similar structure on CCD25-20 show that the charge level in the register falls from saturation to <10 electrons in 100μs.[2] The charge storage capacity of the register is approximately four times that of the pixel.

It was originally anticipated that the bottom corner regions of the image section would be tapered inwards to make room for the charge detection circuitry at the ends of the read-out register, as is the arrangement with other EEV devices such as the CCD15. This is however more difficult to achieve with smaller pixels through the necessary quantisation of layout dimensions, and would require a block of pixels several hundred square to be tapered. It is now found that a more convenient solution is to bend the register channel through 180° such that the detection circuitry can be placed well away from the edge of the chip, thereby avoiding any change of active pixel geometry. A total of 50 blank "run off" elements are included before each detection circuit.

3.4 Detection Circuity

Figure 2 shows a schematic diagram of the output circuitry. This comprises three separately connected electrodes (to be described later) and the charge detection amplifier. The detection node and reset switch are conventional, but the amplifier is an advanced two-stage type now successfully employed on other EEV devices, such as the CCD30 and CCD35. A two-stage circuit enables the first stage transistor to be very small, thereby maximising the responsivity and minimising the noise, with only minimal loading from a much larger second stage transistor which provides a high level of drive capability. Although a.c. coupled, the circuit has d.c. restoration activated by the transfer of charge into the read-out register via an internal connection to the last image section clock phase. It may be noted that the connections to the circuit are identical to those of a single stage type, the only difference being a standing current (<1mA) now flowing in the substrate connection. There is no light emission to cause generation of spurious charge. The external load is non-critical, and can be a resistor of about 5-10kΩ or a constant current of about 2-5mA.

Figure 3 shows the predicted noise performance in comparison with predicted and measured values for EEV CCD30 and CCD35 devices. These characteristic curves show the usual low frequency "floor", a consequence of 1/f noise in the first stage transistor, with the level increasing at higher frequencies through the white noise component

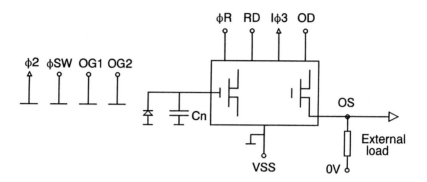

Figure 2   *Schematic diagram of output circuitry*

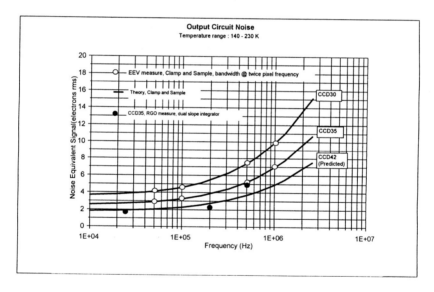

Figure 3   *Predicted noise performance in comparison with predicted and measured values for EEV CCD30 and CCD35 devices*

becoming dominant. Theory[2] shows that, provided the circuitry is reasonably well optimised, at any frequency the noise (in electrons rms) is proportional to the square root of the detection node capacitance, C_n, which is made up of the diode and other parasitic capacitances, or the square root of the reciprocal of the responsivity. The detection node capacitance of the CCD35 is half that of the CCD30 (nominally 25fF as against 50 fF), hence the $\sqrt{2}$ lower noise. As the node capacitance of the CCD42 is estimated to be 12 fF (±15%), another factor 2 lower, a further $\sqrt{2}$ reduction of noise is anticipated.

With the overall voltage gain of the amplifier being typically 0.45 - 0.50, the output responsivity is estimated to be about 6µV/electron. Test devices have demonstrated the reduced node capacitance, but the noise could not be measured because the circuits were of an earlier form with significant light emission.

Back-illuminated devices may be slightly inferior in performance to their front-illuminated equivalents on account of extra node capacitance arising from the dielectric constant of the face-down chip-mounting materials being higher than that of air, thereby increasing the magnitude of the fringing fields between components.

It should be noted that theoretical curves and the EEV measured performance figures shown in Figure 3 are for a relatively conservative pre-sampling bandwidth (twice the clock frequency) and that superior performance should be possible with reducing the bandwidth to nearer the theoretical limit, e.g. using the dual-slope integrator as shown by the RGO measured figures for CCD35.

Figure 4 shows the predicted output settling time constant, again in comparison with CCD30 and CCD35 results. Provided the load capacitance is not too large, the settling time should be sufficiently short to permit operation at frequencies up to at least 1MHz.

The peak charge detection capacity is approximately 300k electrons, as set by the voltage swing at the node being within the 4 volt nominal maximum. In cases where this is insufficient, the output node capacitance can be increased to approximately 36fF by taking the bias on the last register electrode (designated OG2) to about +10V (not critical), thereby increasing the charge capacity by a factor 3 and decreasing the responsivity by the same amount. The noise will also increase by a factor 3, not $\sqrt{3}$ as the size of the first stage transistor remains unchanged, i.e. non-optimised.

3.5 Output Gates

The last three electrodes of the read-out register have separate connections (as indicated in Figure 2), and can be operated in a variety of modes. The first electrode is provided as a "summing well", hence the designation as φSW, but can be clocked

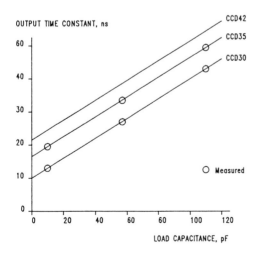

Figure 4 *Output settling time constants*

normally as Rϕ3. The well capacity is approximately 6 times that of the pixel. The second electrode is the normal output gate, designated OG1. The third electrode, designated OG2, is either biased about a volt more positive than OG1, or at a higher level (+10V) to increase the detection node capacitance (as already described).

3.6 Quantum Efficiency

Back face passivation is by boron implantation and laser anneal which shows good stability, though by this process it is more difficult to achieve high QE at ultra-violet wavelengths. Typical measured performance for this process is shown in Figure 5, together with an indication of specified and target QE.

3.7 Edge Spacing

The sides and top edge spacings of the chip (i.e. outside the active area) have been designed as a minimum, bearing in mind the various operational constraints.

As the edge spacing is decreased at the sides, the width of the metal tracks connecting to the polysilicon electrodes must be reduced and, in combination with the electrode capacitance, the consequent increase in series resistance increases the RC time constant and therefore reduces the maximum parallel transfer rate. A track width of 20μm gives a maximum parallel transfer rate of about 20kHz. Using 20μm wide tracks, the outer edge of the peripheral substrate ring (the outer-most feature in the design) is then 160μm from the edge of the first active pixel.

The top edge spacing is designed to be 100μm with a guard ring and 50μm without. The guard ring may be necessary in devices fabricated on thick, high resistivity epitaxy for "deep depletion" applications, but may not be required using normal silicon.

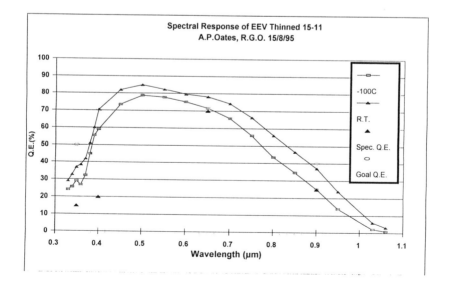

Figure 5

In both cases an additional ≈50μm will be allowed between the edge of the substrate ring and the edge of the chip and the package should extend a further ≈50μm to protect the CCD from edge damage.

4 DEVELOPMENTS

While designing the device to meet the specification requirements, it was decided to include features which would enable further developments.

4.1 Frame Transfer

Provision has been made for the electrodes of the image section to be split into two vertical sections, with appropriate extra clock phase connections (for which bond pads and gate protection circuits are reserved), to realise a frame-transfer variant. This variant would be produced with a simple change of metal pattern, but is somewhat wider than the basic full frame type, adding a further 180μm/side. The maximum transfer rate for a 2048 x 2048 frame transfer device should be about 20kHz, allowing frame transfer to take place in 100ms.

4.2 Anti-blooming

Conventional surface anti-blooming structures result in a significant loss in quantum efficiency. Allowance has been made in the design to support a new "shielded anti-blooming" structure which has been demonstrated to give minimal loss of quantum efficiency in front illuminated CCDs and would lose still less in a back illuminated device. Estimated loss above 900nm is <2%, with no loss below 900nm. In a system with a shutter there will be no image smear during readout so tolerance to optical point overloads will be limited only by optical side effects such as scatter.

4.3 Multiple output

As described above, two stepper fields are required to "stitch" a CCD, a "middle" and an "ends" field. The Ultratech Wide Field stepper reticle has three fields available, so in the spare field we included a more speculative termination which will yield a device of the form shown in Figure 6. With this structure we are attempting to further increase the readout rate by a factor 4 (using outputs 1 to 4, reading to the right), but retain the ability to read all data through one output (output 5, reading to the left), allowing a simpler system with no requirement for gain and offset compensation. In case anyone feels the need, each of these outputs includes a floating gate "skipper" amplifier in addition to the floating diffusion amplifiers. These are also a two stage design with similar settling time, allowing freedom in the choice of optimum skipping frequency.

4.4 Wavefront Sensing CCDs

A small space was still available in a test field so, to continue with the astronomy theme, two CCDs were designed (Figures 7 and 8) for adaptive optics application. These devices will be back illuminated, having a target performance of 1000 frames/sec. with a

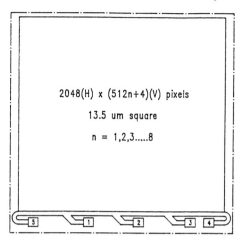

Figure 6

read noise of 5 electrons rms. The output amplifiers are the same as those on CCD42, so we anticipate ≈5 electrons read noise at 1MHz. In order to read 1000 frames/sec. with this noise performance it will be necessary to run at 2MHz pixel rate and reset only once in each line, so allowing the same sampling time as is available at 1 MHz. CCD39A is intended to meet the needs for wavefront sensors on the Gemini 8 metre telescopes.

In order to operate at this frame rate and noise performance, a CCD with more than 80 x 80 pixels will require >4 outputs and therefore >2 outputs on each register. CCD39B is designed to evaluate the necessary structures.

5 DESIGN SUMMARY

Tables 1 and 2 summarise the principal design and performance parameters of the basic CCD42-80 chip variant which are in line with the user requirement. A compromise was made between butting loss and line transfer rate and a novel output circuit with variable node capacitance was designed to handle about 1 million electrons and also maintain a very low noise capability. Devices will be operated with drive pulse sequences and voltage levels similar to those with other EEV large area sensors.

6 PRELIMINARY RESULTS

Since these new devices have literally just come off the production line, it is only possible to present very preliminary results here. Wafers from the first production batch were DC probe tested (at EEV) and one wafer was selected to provide first samples. Several frontside devices were packaged, and then cryogenically cooled and slow-scan tested (at RGO). Three CCD42-10 (2K*512) and one CCD42-80 (2K*4K) devices were tested. Since the CCD42-80 device was of most interest, this device was evaluated the most, and these results are shown below.

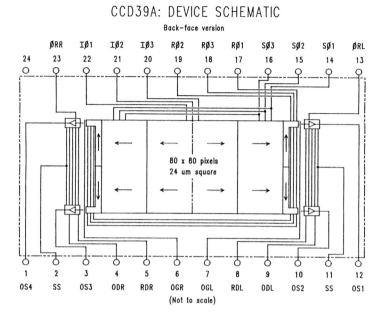

Figure 7

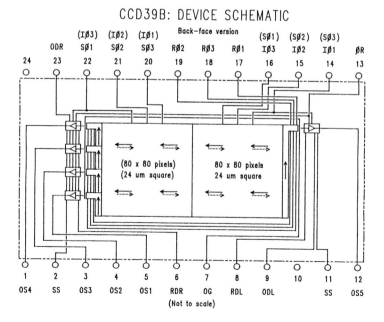

Figure 8

Table 1 *CCD42-80 Design Data : Basic Design Features*

Device format	Full frame (FT as variant)
Number of pixels	2048(H) x 4100(V)
Pixel size	13.5µm square
Operating mode	Normal (IMO as variant)
Inactive edge spacing	
- sides (to VSS ring)	160µm (340µm FT)
- top (to VSS ring)	50µm to 160µm (see text)
Register format	Split, with gated dump drain
Number of blank elements	50 (each end)
Number of output gates	3 (each end)
Number of amplifiers	2

Table 2 *CCD42-80 Design Data : Performance Estimates*

	OG2 low	OG2 high
Charge storage capacities		
- Pixel (electrons)	170k (typ), 150k (min)	
- Register	x4 pixel	
- Summing well	x6 pixel	
Vertical transfer rate (max)	20kHz	
Horizontal transfer rate	>1MHz	
Output circuit performance	OG2 low	OG2 high
- detection node capacitance (fF)	12	36
- overall responsivity (µV/electron)	6	2
- noise at 50 kHz (electrons rms)	1.8	5.4
- noise at 1 MHz (electrons rms)	5.0	15
- settling time constant (ns, 10 pF load)	25	25

6.1 Operating Temperature

-115°C.

6.2 Cosmetics

Very good. No column defects were evident, and only two partially blocked columns could be seen. No extensive search for very low level pixel defects has been made yet.

6.3 C.T.E.

The devices produced satisfactory first images, but serial and parallel CTE remain to be quantified. Other EEV (05-xx series) devices have shown excellent CTE; it may be that these new devices require some optimisation/adjustment for optimal performance.

6.4 Dump Drain

These devices have a gated dump drain to allow very rapid discharge of the serial register, for rapid clearing of the array. We have not yet tested this function fully, but confirmed that with the gate 'Off' (ie LOW) the array operates normally; with the gate 'on' (HI) charge is dumped. We expect this feature to be very useful in enabling very rapid clearing of the array, and also enabling more rapid skipping of rows during windowed readouts.

6.5 Switched Output Gates

The output nodes have a double output gate; modulation of a control voltage on OG2 changes the node capacitance, output sensitivity (and noise). Results taken (see below) indicate that the node sensitivity is switched by approximately a factor of two (cf design factor of three); to be confirmed on other devices. We also expect this feature to be very useful, to allow optimised use in different signal level regimes.

6.6 Readout Noise

These devices have been designed to provide very low noise, especially at higher readout rates. The tables below tabulate our results for both states of the switched gate (OG2); the upper table has the highest sensitivity, and lowest noise. Since the measurements are very preliminary, we have chosen to present only the basic, 'raw' data here. The effective sampling frequency was changed by using several values of double-correlator sampling (DCS) times, although our overall pixel duration was typically 15-50µS. Use of short sample times, allowed us to measure noise at relatively high readout rates. The DCS circuit uses a double sample, for reset and signal measurement.

Very low readout noise performance has been achieved, especially at the 4µS sampling times. At lower frequencies the anticipated reduction of (white) noise with frequency is not seen - this apparent excess 1/f noise is under investigation. However the baseline noise figures are still excellent! At the higher frequencies, we also see excellent low noise - although some caution is necessary since we did not optimise our system gain for these measurements. The system noise component has been subtracted in the table, but can be seen to be finite.

6.7 General Comments

All devices tested showed broadly consistent performance. Once the initial control voltages, and clock sequences were established, device behaviour was repeatable and consistent between devices. Indications are that sensitivity and noise do not vary by more than 10% or so from one output to another.

Table 3 *Readout Noise, at various sampling times.*

On-chip (double)amplifier gain ~0.4(±30%)

For OG2 = LO   Output sensitivity (at OS) ~ 5μV/e (±50%)

OG2 = LO (maximum node sensitivity)

	(1)	(2)	(1)-(2)	3	$(1-2)^{1/2}$ *3
DCS Time Constant	Total Variance (CCD+System)	System Noise Variance (no CCD)	Net Variance (detector noise)	Conversion factor	Derived Readout Noise
T(μS)	(ADU)	(ADU)	(ADU)	e⁻/ADU	e⁻ rms
0.5+0.5	2.4	?	2.5	2.9	4
1+1	4.5	2.5	2	2.4	3.5
4+4	6.5	3.5	3	0.94	1.6
10+10	14	3.5	10	0.6	1.9
20+20	19	4	15	0.5	2.0

OG2 = HI (minimum node sensitivity)

	(1)	(2)	(1)-(2)	3	
DCS Time constant	Total variance (CCD+system)	System noise variance (no CCD)	Net variance (detector noise)	Conversion factor	Derived Readout Noise
TμS	(ADU)	(ADU)	(ADU)	e-/ADU	e⁻ rms
1+1	3.4	2.5	1	4.6	4.5
4+4	4.5	3.5	1	1.8	1.8
10+10	12	3.5	8	1.2	3.6
20+20	18	4	14	1.0	4

Since these devices are frontside CCDs, we have not yet measured their spectral response; final thinned devices will be measured fully in the very near future.

We must emphasise that these are very preliminary measurements of first samples of this new device. Many aspects of operation and performance remain to be fully examined. However we are very encouraged by test results so far, and indications are that many of the design goals have been met.

References

1. P J Pool, S R Bowring, W A Suske, and J E U Ashton, Proc. SPIE Vol 1242, 1990.
2. T Eaton : private communication regarding device assessment for the MERIS instrument on the European Space Agency Satellite ENVISAT 1.
3. D J Burt, CCD Performance Limitations: Theory and Practice Nuclear Instruments and Methods in Physics Research. Vol A305, 1991, 564.

A CIRCULAR CCD FOR APPLICATION WITH A FABRY-PEROT INTERFEROMETER

Wayne W. Frame
Ball Aerospace and Technologies Corporation
Boulder, Colorado 80306

T. L. Killeen
Space Physics Research Laboratory
University of Michigan
Ann Arbor, Michigan

1 INTRODUCTION

Remote-sensing measurements of atmospheric winds, temperatures and densities have been made successfully using Fabry-Perot interferometers (FPIs) on board the Dynamics Explorer-2 (DE-2) satellite and the Upper Atmosphere Research Satellite (UARS). Variants of these successful instruments, developed at the University of Michigan (UM), have now been selected for the TIMED mission and have been proposed for NASA Discovery, SMEX, and MIDEX missions. They will also be proposed for the upcoming Earth Probes opportunities. An FPI is being considered for use in conjunction with a spaceborne Lidar system to measure global tropospheric winds. In addition to the use of the FPI technique for spaceborne studies, there are about 30 ground-based facilities worldwide that employ the FPI technique for the routine measurement of upper atmospheric winds and temperatures.

The broad relevance of optical interferometry in research programs may be attributed to the unique advantages offered by the technique for the direct remote sensing of the dynamics (winds) of planetary atmospheres. The Space Physics and Solar System Exploration Divisions of the Office of Space Sciences (OSS) and the Earth Science Programs of the Office of Mission to Planet Earth (OMTPE) all have requirements for knowledge of atmospheric winds, and it is to be anticipated that optical interferometers will continue to play a significant role in NASA missions for the foreseeable future.

While the optical interferometric technique offers tremendous advantages in terms of spatial coverage and high accuracy, it remains a technical challenge for two main reasons. Firstly, wind-measuring interferometers require very high spectral resolution and rigorous off-band rejection. Secondly, spaceborne FPIs require high photometric sensitivity, particularly when viewing dim night-time atmospheric emissions (10's-100's Rayleighs only). In order to address this latter challenge two approaches may be taken: (1) incorporate a large instrument aperture (with all the consequent penalties in weight, power and cost) and (2) increase the FPI detector sensitivity while retaining a relatively small instrument aperture. This paper describes a recent breakthrough in technology related to the second of these two approaches - namely, the development of a concentric-ring, circular, Charged Coupled Device (CCD) detector, with custom geometry optimized for the detection of FPI interference ring patterns. The circular device combines the intrinsically high sensitivity of bare (unintensified) CCDs with extremely low noise and distortion-free charge transfer efficiency.

1.1 Science and Technology Background

The circular CCD development is an outgrowth of previous detection technology developments for space flight. The following paragraphs describe some of the key prior work and motivation for the present development effort.

Fabry-Perot interferometers, which view naturally-occurring atmospheric emission lines, produce interference ring patterns in the focal plane. These interference patterns have circular symmetry, with each integral order producing a bright ring (or fringe) of equal area (see Figure 1). Such patterns (in emission or absorption) provide a rich source of geophysical information [*Killeen et al.*, 1984]. The exact radius of each ring is related to the bulk motion (winds) in the emitting atmosphere, the width of each ring is related to the temperature, and the brightness is related to the density of the emitting species.

Various methods have been used to extract the geophysical information from such observations [see *Hernandez and Killeen*, 1986, for a historical perspective]. Conventionally, pinhole photomultiplier detectors have been combined with scanning techniques to systematically move one fringe across a central location. This technique has the drawback that a lot of time is wasted on the dark portions of the interference pattern - regions that contain little useful information. It has the further disadvantage of requiring some type of scanning mechanism (an expensive requirement in space applications).

In the late 1970's and early 1980's an improved approach was developed in a partnership between the Space Physics Research Laboratory of the University of Michigan and ITT, based on an array detection scheme. The new image plane detector (IPD) was flown on the DE-2/FPI and UARS/HRDI instruments. The IPD was based

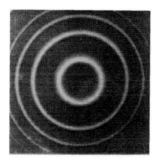

Figure 1 *Interference ring pattern in focal plane of an FPI. This pattern shows ⁻3.5 orders.*

on an S20 photocathode, a stacked microchannel plate electron multiplication stage and a fixed concentric ring anode structure [*Killeen et al.*, 1983]. This detector eliminated the need to spectrally scan the instrument. This approach, however, allowed only one order at a time to be imaged, limiting the total amount of detected light. It was also limited by the relatively modest quantum efficiency of the S20 photocathode (˜5% at 6300-Å).

Both the DE-2 and the UARS/HRDI instruments were highly successful. The single etalon instrument built and flown by us on DE-2 successfully charted the dynamics of the Earth's thermosphere and provided the essential space qualification for the FPI technique. The DE-FPI technique has been presented in a series of papers [*Hays et al.*, 1981, 1984; *Killeen et al.*, 1981, 1982, 1983; *Killeen and Hays*, 1984] wherein various aspects of the instrumentation are discussed. Analysis of data from DE-FPI has led to the publication of over 70 scientific papers (see review by *Killeen and Roble* [1988] for a comprehensive reference list).

In addition to the successful flight of the DE-FPI, a triple etalon interferometer, the High Resolution Doppler Imager (HRDI), has been built at the SPRL and is currently flying on the Upper Atmosphere Research Satellite (UARS). The UARS/HRDI instrument enables absorption features in the Troposphere and Stratosphere to be observed at high spectral resolution, leading to measurements of the wind fields in the Earth's lower atmosphere.

Recently, "bare" or unintensified CCDs have emerged as the detector of choice for the ground-based FPI community, owing to the much greater quantum efficiency, imaging capability and ruggedness. While CCDs are in regular use for ground-based interferometry [e.g., *Niciejewski et al.*, 1992], they have not yet been used routinely for spaceborne instrumentation. One drawback of the conventional CCD for Fabry-Perot interferometry is the large number of pixels that need to be read out (with consequent large integrated read noise) and the intensive image processing needed to convert the rectangular format into the circular format.

The viable performance of conventional (rectangular-geometry) CCDs in ground-based interferometry is indicated by Figure 2, which shows measurements of the mesospheric (˜87 km altitude) winds over Ann Arbor using a prototype FPI designed for flight in the NASA planetary program. The work with bare CCDs demonstrated the gain in sensitivity and the extended spectral range achievable with compact CCD-based Fabry-Perot for planetary studies. The accuracy of the ground-based observations shown in Figure 2, made with a CCD detector, are ˜10 m/sec, which is about a factor of 10 improvement over earlier results using conventional GaAs photo multiplier tube technology.

The approach using a conventional rectangular CCD has some limitations. First, the rectangular image has to be converted to the circular (equal-area annuli) geometry in order to assess the Doppler characteristics of the emission line. This conversion, which involves centroiding and reduction to an equal-area annuli system, is relatively expensive in computer time. Also, the large number of individual (rectangular) pixels or super pixels that are needed to retain spatial fidelity for the outer orders in this figure limit this approach to relatively few orders of interference (up to ˜3). Another problem involves the effects of read noise (˜5 electrons per pixel read) that becomes serious due to the multiple pixel-reads (over 15,000) needed to fully characterize the fringe pattern with rectangular pixels.

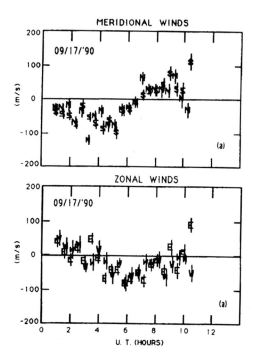

Figure 2 *Wind measurements at ˜87 km altitude made using bare (conventional)*
CCD detection from a ground-based observatory in Ann Arbor, Michigan

1.2 Circular CCD

In this paper we describe an alternative approach to FPI detection through the use of a highly innovative detection scheme. The approach taken is an intuitively simple, but far-reaching one. We have developed a device that *customizes* the pixel geometry of a CCD (see Figure 3), optimizing it for the particular interferometric application and simultaneously reducing the total number of pixels by a large factor. We estimate that the detector approach proposed here will lead to an inherent increase in sensitivity of a factor of ˜100 over the HRDI/UARS spaceborne FPI. Such an improvement will make future FPIs much smaller and of lower cost.

2 PHYSICAL DESCRIPTION OF DETECTOR

The detector is comprised of eighty equal area concentric rings or annuli (Figure 4). Each annuli has an area of 2.209 square mm. The diameter of the photo active area is 1.5 cm.   The central area is circular with a diameter of 1677 microns. The width of the 2nd ring is 341.5 microns. The width of the 80th ring is 41 microns. Each ring is divided into 4 pixels to facilitate readout. Each ring, with the exception of the central

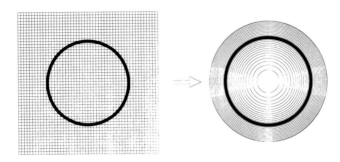

Figure 3 *Schematic illustration of the customization of the pixel geometry of a CCD to match the geometry of the equal-area, concentric-ring interference pattern*

area, has some obscuration. This obscuration varies from 14.2% for the 80th ring to 1.32% for the second ring and zero for the central area. There is a serial readout register on each side of the detector. The pitch of the readout register is 21 microns.

The readout register is a conventional three phase design. There are 160 cells plus eight extended cells in each register. Two pixels from each ring feed two serial readout register cells, lower and upper. The 80th ring feeds the first and last readout register cell; cell 9, (8+1) and cell 168, (8+160) respectively. Each half of the quadricated central area feeds the 88th and 89th cells, etc. The eight extended cells serve their usual role of gaining physical space to allow mating to an on-chip charge detection amplifier and getting everything electrically connected.

Between each annuli there is a 6 micron wide channel stop. At the vertical center line separating left and right, a channel stop forms a barrier dividing the ring into two halves. At the horizontal center line there is a further division, thus resulting in each ring being divided into four pixels. At the horizontal division each pixel is routed out to the side and mated with the readout register cell. Outside the circular area of the detector a light shield metalization is employed to keep photons from impinging on silicon except where desired. This light shield is a second level metalization, the first being the electrical hook-up level. Over the horizontal extension of each pixel, light shield metalization is employed to prevent spatial contamination (Figures 5a and 5b). Each of these 21 micron wide horizontal extensions is mated to a readout register cell as described above. There is a mini-channel or notch in each active channel and its attendant horizontal extension. All rings are wider than these 21 micron wide horizontal extensions. Hence, signal charge experiences a narrowing channel as it is transported to the output. The mini-channel is of constant width which mitigates this

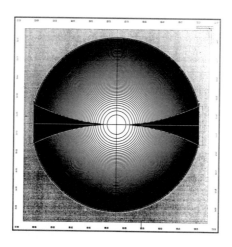

Figure 4 *80 ring CCD outline*

situation somewhat. Furthermore, even at high signal levels, charge density is very low in terms of electrons per square micron. There are two gates that separate the terminus of each horizontal extension and the readout register, a collector gate and a transfer gate. The function of each gate is described below.

Inboard of the collector gates, left and right, the photo active area of the detector is covered by a conventional three phase clocking structure. This structure has left/right mirror symmetry to generate simultaneous left/right charge transport. The outer ring is surrounded by a drain to limit edge effects on the 80th ring.

The 80 ring circular device has very much larger pixels than a conventional device. Therefore, full well capacity is not set by the imaging pixel but by the readout structure. The area of each pixel is 0.55 square mm. Therefore, dark current is an important consideration. In many applications multi-pinned phase (MPP) mode of operation is used to suppress the surface state dark current generation rate. At the silicon, silicon dioxide interface, the crystalline lattice is discontinuous. This results in a much higher concentration of dark current generation centers than in the bulk. In a conventional CCD charge packet separation is maintained in the MPP mode by an implant which builds an electric field into the silicon. In the MPP mode of operation the entire surface is inverted, thus an excess of holes exists at the surface. These excess holes annihilate thermally generated electrons by re-combination as they are created. The surface is inverted by taking all clocking electrodes sufficiently low. The intended use of the circular detector requires radial resolution, not angular. Therefore, the entire signal integration area can be inverted with no implant and with no loss of information.

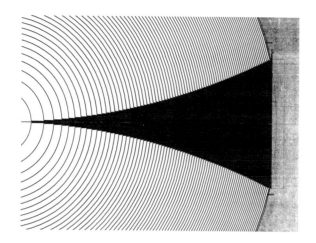

Figure 5a *Detail of light shield*

The transfer gate is also set to a low level to create a barrier between the readout register and signal integration area so that the start of exposure N+1 can occur while exposure N is being read out. The collector gate, which is inboard of the transfer gate, serves as a charge collection site. Thus, surface state dark current suppression can be achieved with the circular detector without the need of a special barrier implant as is used in devices operating in the conventional MPP mode.

If a dark current noise budget of 5 electrons RMS is allowed for in a one-second exposure, then the dark current rate must be no more than 25 electrons per second in the 0.55 mm^2 pixel. This corresponds to a dark current rate of 7.27E-7 nanoamps/cm^2. This level of dark current requires an operating temperature on the order of -65 °C or lower for a one-second frame rate or -78 °C for a ten-second exposure.

After the signal integration period, collected signal charge is transferred to the readout register. It is desirable to accomplish this task as quickly as possible in order to

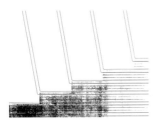

Figure 5b *Light shield and channels at horizontal extensions; mini-channel also shown*

achieve high temporal resolution. Signal charge is swept from the rings to the collector gate by manipulating the clocking electrodes, P1, P2 and P3. These clocking electrodes are analogous to the parallel clocks of a conventional rectilinear CCD. The clocking electrodes are mirror symmetric about the vertical center line so that signal to the right or left of the center line is swept in the direction appropriate for readout. The pitch of the three phase clock set is 27 microns, 9 microns per electrode. At the vertical center line the action of the clocking electrodes produce field gradients in line with the desired charge transport direction. However, as signal charge is *pushed* around the arc, the clocking field gradients and the needed transport direction are tending towards right angles. This situation is mitigated by having the mini-channel slanted towards the outside, near the end of arc section of the channel; see Figures 5b and 5c. Figure 5b shows channels near the outer edge, Figure 5c near the center. The result is that for the worst case situation, the 80th ring, charge transport has to travel 31 microns per clock electrode width, or 3.44 microns in the y-axis per micron in the x-axis. This means that the clock time must be slow enough to allow time for charge diffusion. The clock set cycle time used for testing was 144 μS. For the intended application, illumination along the length of any particular pixel should be fairly uniform. That is, intensity will be varying radially but not angularly. With this being the case, the pixel physical size is so great that signal charge can easily be totally contained in the mini-channel at signals much greater than 200K electrons per pixel. Hence, the readout register will limit signal charge before signal spills from the mini-channel.

There are 270 P_n clock sets on each half, left and right side, of the device. It is a good idea to over clock; we tested with 300 P_n clock sets. Hence, a sweep time of 300 x 144 uS = 43.2 milli-seconds is required. This time can be thought of as the temporal *smear* time. There was very low residual signal in a second read of the device at all signal levels (< 2%). The parallel clocking can be executed at a faster rate with some increase in residual signal. This may be a valid system trade in the

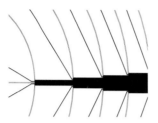

Figure 5c *Light shield and channels near center of device*

future. The serial register was clocked at a rate of 24 µS/pixel. The parallel clocking sequence must be repeated a minimum of 270 times to clear the imaging area of signal charge.

3 TEST SET-UP DESCRIPTION

The circular CCDs were tested in Ball Aerospace's CCD laboratory. Figure 8 illustrates the test set-up. This facility uses a time base generator and bias supply built by Pulse Instruments. Timing code is written and compiled on a PC. The device under test is placed in a vacuum dewar and cooled by liquid nitrogen. A heater control servo is used to set the desired operating temperature. Preamplifiers with a voltage gain of 13.4X are co-located with the CCD inside the dewar. The output signal is processed by a correlated double sampling (CDS) circuit to suppress the reset and 1/F noise of the on-chip charge detection amplifier. The CDS electronics has selectable gains of: 1X, 2X, 5X, 10X and 20X. The signal is digitized to 16 bits and passed to a PC via a Datel, Inc. data acquisition board. IDL is used to manipulate and process test data. A mechanical shutter was used to control exposure.

4 TEST RESULTS

4.1 Performance Parameters

The nominal on-chip charge detection amplifier responsivity is 1.5 ±0.2 µV/e-.

The amplifier responsivity was measured by use of an iron 55 source. Because of the very large pixels in the circular device, the standard statistical measurement

Figure 6 *Picture of wafer with six Ball Aerospace designed custom CCDs, from foundry at EG&G Reticon, Sunnyvale, California*

Figure 7 *Picture of a packaged circular CCD*

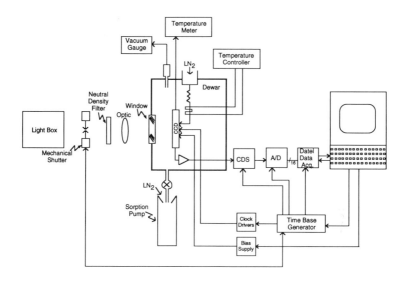

Figure 8 *Schematic diagram of test setup*

procedure could not be used. Serial CTE on a rectilinear device from the same production lot was measured and was found to be greater than the test setup limit of .9998 over a shift of 3800 pixels.

The nominal readout noise was measured at ≤ 10 electrons RMS.

Dark current: 1.2 E-15 A/cm^2 @ - 61 °C.

Large signal handling capacity: >200K electrons/pixel.

A rough quantum efficiency (QE) measurement was executed on device S/N 1. A UDT Instruments optometer, model No. 370, and an Ealing narrow band filter was used for this assessment. The center frequency of this filter is 650 nM and the FWHM is 11.3 nM. The measured QE is 57% at a temperature of -80 °C. This is the QE for the active area; it does not take account of any loss due to the light shield *fill-factor* in any particular ring. For example, the anticipated QE of the 80th ring is 57%(1 - 14.2%) = 49%. There are significant error bars associated with the QE measurement as a result of the rather crude test set up.

4.2 Imaging Results

Following are *snap shots* of some test data taken with flat field illumination. The y-axis is digital number (DN) of the A/D conversion. Full scale is $2^{16} = 65,536$ which

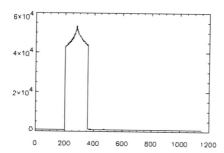

Figure 9a *Average signal 170K e-, temp = -76 °C, S/N 12, left output*

corresponds to a signal level of 4 volts. The x-axis is pixel address location. There are six readouts of the serial register shown. Each readout of the serial register is 192 'pixels' in length: 8 extended pixels, 160 data pixels followed by 24 over clocked pixels. (The over clocked pixels do not map into real physical pixels; they are used for baseline information. There is no special significance to the number 24.) The first readout is before the parallel shift of integrated signal into the serial register. This readout is used to purge the serial readout register. The next readout is the actual data (integrated signal). The actual data read is followed by a sequence of four more parallel shifts and subsequent serial register readouts for a total of 6 X 192 = 1142 'pixel addresses' delineated on the x-axis. Ideally, there would be no signal in the third and subsequent readouts since the shutter in the optical path was closed before the first read out.

The *pagoda roof* shape of the flat field video waveform (see Figure 9) is a result of the progressive percentage obscuration of the rings, center to outer edge (0% to 14.2%). With the exception of the center section, a constant 21 microns is added to the obscuration per ring, center-to-edge. Since ring width is a variable, the active area loss is not a linear function with ring position, varying faster near the center than near the edge. This circumstance gives rise to the *pagoda roof* shape. The baseline tilt visible at low signal levels is a result of the test set electronics and not the circular CCD.

Exposure times for flat fields were 1 second. For dark current measurements the integration time was 20 seconds and 1 minute.

5 ADVANTAGES OF A CIRCULAR DEVICE COMPARED WITH A RECTILINEAR DEVICE

For the sake of comparing this device to a conventional rectilinear device with 21 micron square pixels. The area of a pixel of the circular device is > 1200 times that of

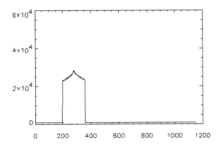

Figure 9b *Average signal 92K e-, temp=-80 °C, S/N 12, left output*

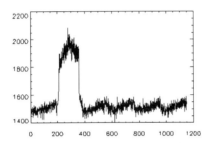

Figure 9c *Signal level 180 e-, CDS gain=10X, temp=-70 °C, S/N12, right output*

the rectilinear device. The Circular device operates well at signal levels of 100 electrons/pixel and lower. This is equivalent to < 0.1 electron/pixel of the conventional device.

There are significant advantages in charge transfer efficiency (CTE) of the circular device for Fabry-Perot applications. In a conventional rectilinear device, less than ideal parallel CTE not only results in lost signal, but more importantly in spatially displaced information due to deferred charge. In the circular device deferred charge does not cause any spatial misinformation; at worst the information is only temporally displaced. In the serial or readout register axis spatial skewing can be mitigated with a circular device with four outputs so that the pixels each ring experience the same number of shifts.

Post process of the signal from a circular device is clearly a much easier task as a result of the detector geometry that fits the problem. The central area of the circular CCD can function as a quad-cell as its geometry is identical. This feature has the potential to assist in the optical alignment of an instrument.

6 COSMIC RAY SUSCEPTIBILITY

The circular CCD is a fairly large area device, each half being nearly a square centimeter (0.88 cm^2) in area. With such large pixels and large overall area, the device makes a very nice cosmic ray detector. The cosmic ray flux at our location was observed to be on the order of 4 per minute per cm^2. This was not a scientific measurement but a rather   coarse observation. The observation was made in late afternoon in September. Elevation of Boulder, Colorado is 5400 feet above sea level. Figures 9 and 10 delineate the capture of such an event.

When the concentric ring fringe patterns are centered, cosmic rays may be discriminated against by comparing amplitudes of each ring set of four pixels. These amplitudes should agree fairly closely. An absorbed cosmic ray in any one of the four pixels will raise its amplitude. The high valued pixel can be rejected and the average of the other three substituted.

7 CONCLUSION

The circular CCD development has been successful. Matching the geometry of a detector to the problem offers a substantial improvement in signal-to-noise ratio performance as well as greatly reduced post signal processing. The area of a pixel, one-fourth of a ring, is more than three orders of magnitude greater than a typical rectilinear CCD. This has the potential to greatly increase signal-to-noise ratio at very low light levels. As can be seen in the test data, charge transport is maintained at extremely low signal levels, equivalent to level < 0.1 electrons per pixel of a typical rectilinear device. Since pixel area is great the device must be operated cold in order to suppress dark current to acceptable levels, even for fairly fast frame rates. However, the cold operating temperature is still reasonable and readily achievable. It was

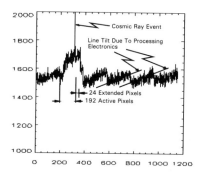

Figure 10 *Signal level 34 e-; Cosmic event captured at pixel readout address location 312*

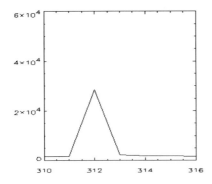

Figure 11 *Cosmic ray event with magnified x-axis scale and expanded y-axis*

observed during testing that charge transport was improved at cold temperatures. A possible explanation is that electron mobility increases at cold temperatures. Furthermore, the mean free path is increased which may tend to make it easier for signal charge to find its way to the mini-channel. The recommended operating temperature is $\leq 80\ °C$.

The intended operating mode is binned; that is, the entire quarter ring is considered as one pixel. However, the device can also function to some level as an area array. Indeed, the device was first tested in this mode. The clock electrodes are of constant pitch horizontally. However, the rings are all of different width, as needed to create equal areas. Thus, when the device is used as a two dimensional array, the pixels in any quadrant are all of different areas.

References

1. P. B. Hays, T. L. Killeen, and B. C. Kennedy, *Space Sci. Instrum.*, 1981, **5**, 395.
2. P. B. Hays, T. L. Killeen, N. W. Spencer, L. E. Wharton, R. G. Roble, B E. Emery, T. J. Fuller-Rowell, D. Rees, L. A. Frank, and J. D. Craven, *J. Geophys. Res.*, 1984, **89**, 5597.
3. P. B. Hays, D. L. Wu, and the HRDI Science Team, *J. Atmos. Sci.*, 1994, **51**, 3077.
4. G. Hernandez and T. L. Killeen, Chapter 5, COSPAR International Reference Atmosphere: 1986, Part I: Thermosphere models, *Adv. Space Res.*, **8**, No. 5-6, pp 149-214, Pergamon Press, 1988.
5. T. L. Killeen, P. B. Hays, B. C. Kennedy and D. Rees, *Appl. Optics*, 1982, **21**, 3903.
6. T. L. Killeen, B. C. Kennedy, P. B. Hays, D. A. Symanow, and D. H. Ceckowski, *Appl. Optics*, 1983, **22**, 3503.
7. T. L. Killeen and P. B. Hays, *Appl. Optics*, 1984, **23**, 612.
8. T. L. Killeen and R. G. Roble, *Rev. Geophys.*, 1988, **26**, 329.
9. R. J. Niciejewski, T. L. Killeen, and M. Turnbull, *SPIE*, 1992, **1745**, 164.

A PROGRESSIVE SCAN CCD IMAGE SENSOR FOR HIGH SPEED PHOTOGRAPHY

Stephen J. Strunk and Rusty Winzenread

EG&G RETICON
345 Potrero Ave.
Sunnyvale, California 94806 USA

1 INTRODUCTION

Solid state image sensors are beginning to replace film for image capture and motion analysis of high speed events. In addition to inherent advantages over film, such as resolution in low contrast conditions and elimination of complex mechanics, the electro-optical system offers an overwhelming advantage with its concomitant capability for near real-time viewing and analysis.

Designing an image sensor that has performance characteristics suitable for high speed photography presents a significant challenge. The device must have a high spatial resolution and a dynamic range of at least 10 bits. It must operate with full performance at frame rates faster than 1 kHz. Photosensitivity to normal daylight illumination is necessary, as is blooming control for over-illumination of up to 1000 times saturation. An electro-optical shutter, which can set exposure times to less than 0.0001s, is a desired feature to minimize blur of ultra-fast events. Additionally, the imager must be practical from the standpoint of operating complexity, including a minimum number of outputs, modest power dissipation, and low dark current.

The authors have developed a second generation CCD imager with a resolution of 512(H) by 512(V). The device is based on an interline transfer architecture with progressive scan readout and features a vertical overflow drain (VOD) and 2-phase clocking throughout. This new approach offers significant advantages over the previous design, including short exposure capability, enhanced blue response, and reduced dark current, smear, power dissipation and drive complexity. This paper describes essential design details of the array and presents results from our initial evaluation.

2 DEVICE ARCHITECTURE

The image array, illustrated in Figure 1, is implemented as an interline transfer (ILT) design with a split readout mode. Sensing elements are constructed in an array of 512 columns by 512 rows which are split at the center into an upper and lower section. Associated with each photo-element is a vertical shift register stage (VCCD) which allows progressive scan readout. *Progressive scan* refers to an architecture in which full spatial resolution is achieved in a single field, in contrast to an interlaced design which requires

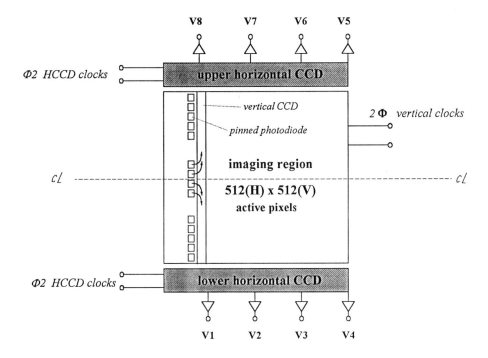

Figure 1 *Schematic illustration of the proposed fast framing CCD image sensor, based on the frame interline transfer architecture. The vertical CCD shift registers are optically shielded to minimize smear during readout*

two readout fields. After integration, charge from the pixel is rapidly transferred into the optically isolated vertical CCD to minimize blur. Immediately after transfer, the pixel begins integration of a new frame while the previous frame is read out.

The pixel-VCCD transfer gate is incorporated within phase 2 of the VCCD and is formed by a threshold adjust implant. The dual purpose electrode permits tighter pixel spacing and fewer control signals but requires a tri-level wave form for operation. Vertical CCD phases are driven at clock rails of -7V and 0V for vertical shifting and +10V for pixel-to-VCCD transfer. After the charge transfer from the photodiode to the vertical CCD, the upper vertical CCDs will shift the charges to the top horizontal readout CCD while the lower vertical CCDs shift the charges to the bottom horizontal readout CCD. Horizontal CCDs are tapped every 64 stages. Both vertical and horizontal registers have been implemented in a two-phase configuration. Each output tap has a buffered charge integrator to convert the signal charge into an output voltage. The integrator is a minimum geometry n+ diode with an adjacent switching MOS transistor to reset the pixel after each data read. The voltage signal of the floating diffusion is buffered by a three stage source-follower amplifier with the ability to drive an off-chip load of at least 8 pF without degrading bandwidth.

3 DESIGN CONSIDERATIONS

3.1 Photo element Design with Vertical Overflow Drain

The photo-element is based on a pinned photodiode (PPD) sensing structure, rather than a conventional polygate or n+ photodiode element. The photogate (CCD) has poor blue response and is insensitive to light below 450 nm. Photodiodes respond into the deep UV, as they do not require an overlying, highly absorbing polysilicon layer for operation. However, conventional n+/p photodiodes suffer from image lag, which appears in the output signal as frame-frame smear. The pinned photodiode overcomes this problem by replacing the n+ diode by a lower doped n-type diffusion, to allow complete depletion of signal electrons before the subthreshold condition is reached.[5] The surface potential of lightly doped n region is pinned to the substrate voltage using a shallow p+ layer, implanted above the diode buried channel.

There are various ways to achieve blooming and exposure control. One way is using an overflow drain and barrier positioned beside the photosensitive area to suppress the blooming. Another way is using a clocking blooming control gate to allow the excess electronic charges to recombine with holes at the surface. Both of these schemes are undesirable from the standpoint of pixel pitch and dynamic range. Reticon has selected the vertical overflow drain positioned underneath, rather than beside, the photodiode based on its efficiency in pixel geometry. A unit cell cross-sectional view of the pixel and VOD is shown in Figure 2.

The pixel is constructed in an n-type material of 10-15 Ω•cm resistivity. A detector p-well is formed in the n-region by implanting boron, followed by a substantial thermal drive, to form a deep junction. Dose and depth of the pixel p-well are important

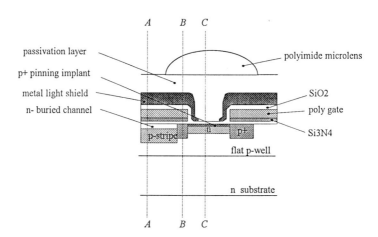

Figure 2 *Cross-section of the pinned photodiode pixel structure. A compact anti-blooming and exposure control feature is implemented by use of the p-well on the n-type substrate. With this structure, excess charge from over-illumination is drained through the pn junction of the substrate; the photodiode is reset by pulsing the n-type to a high positive potential. Section A-A is through the vertical CCD, section B-B is cut through the transfer gate, and section C-C cuts through the photodiode region.*

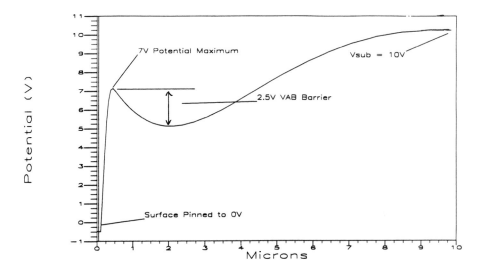

Figure 3 *Two-dimensional modeling of the pixel basic cell showing potential profile into the substrate. A potential barrier with a maximum at 2 microns controls the onset of anti-blooming into the n-type substrate.*

parameters that determine spectral response, storage capacity, and anti-blooming and exposure control performance. A second well is formed in the region of the vertical CCD to provide good isolation of diffusion charge between the pixel and VCCD. This implant is responsible for minimizing diffusion charge from the pixel to the CCD, determining charge transfer efficiency from the diode to the CCD, and maintaining charge transfer efficiency through the vertical register. It is also optimized such that operation of the exposure control feature will not perturb charge within the CCD. This enables the array to operate at 1000 frames per second with a continuous selection of exposure times from 1 ms to 10 μs.

3.2 CCD Design

Results of two-dimensional device simulations are illustrated in Figures 3-5. Figure 4 illustrates the CCD channel potential for barrier and storage phases of the horizontal and vertical CCDs. Note that in contrast to the pixel diode region, Figure 3, the CCD p-well is not depleted through to the n substrate, thus anti-blooming will only occur in the region of the diode. The CCD p-well also remains undepleted during exposure control pulse and therefore signal charge within the CCDs will not be corrupted during the large amplitude pixel reset.

Particular attention was made to the design of the pixel transfer gate. A common pitfall in the interline architecture is the development of a potential pocket or barrier within the transfer region; with the pinned photodiode picture element, image lag is usually due to the presence of this type of charge trap. Results of the device level

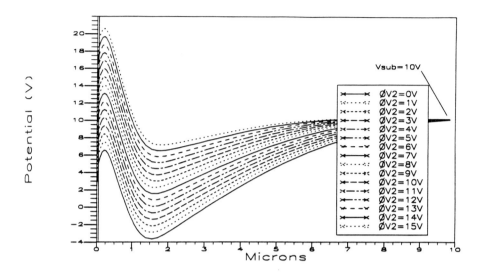

Figure 4 *Potential profile of the vertical CCD region into the silicon substrate over the operational range of bias conditions.*

simulations are shown in Figure 5, indicating the presence of a minor barrier with the transfer gate held to +9V. The barrier is fringed out at +10V and the array exhibits a pixel-VCCD transfer efficiency of over 99 %.

3.3 High Speed Output Amplifier Design

The horizontal CCDs are tapped to provide 8 parallel outputs. Each output is buffered by a wide band buffer configured as a three-stage follower. The -3 dB bandwidth is approximately 150 MHz to allow data rates of over 40 MHz per channel. Arsenic source drain regions are used to produce shallow junctions with minimal lateral spread, improving frequency response by minimizing gate-drain overlap. The high frequency response of the source follower is determined by the dominant pole associated with the output impedance and output driving capacitance. The output impedance is approximately equal to g_m^{-1}, where g_m is the transconductance of the output transistor, and C_L includes parasitic capacitance at the output node. This limiting frequency is as follows:

$$f_H \approx \frac{1}{2\pi} \cdot \frac{g_m}{C_L} \tag{1}$$

Measurement of the small signal gain is illustrated in Figure 6, with an off-chip capacitive load of 8 pF; the measured data is shown together with the SPICE design model.

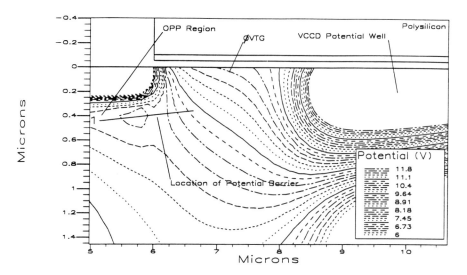

Figure 5 *Two dimensional modeling of the pixel-to-VCCD transfer region illustration potential contours during pixel transfer. The marginal barrier formed at this interface ultimately limits transfer efficiency of this single transfer step, which is manifest as frame-to-frame image lag.*

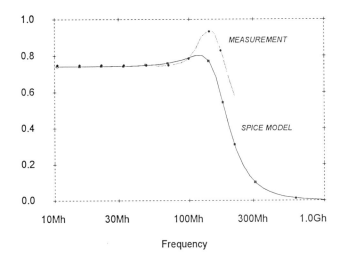

Figure 6 *Model and measurement of the output buffer small signal bandwidth with loads 10 pF (SPICE MODEL) and estimated 8 pF of-chip load (MEASUREMENT). Design requirements call for a half-power bandwidth of 150 MHz to handle the anticipated output data rate of 40 MHz.*

4 PERFORMANCE

4.1 Noise Characteristics and Dynamic Range

In high speed operation, the device read noise is dominated by contributions from the output structures. These have two basic components: (1) MOSFET noise.[6] from the buffer amplifier and (2) kTC-reset noise at the floating diffusion. The latter is included here because the high output data precludes use of correlated double sampling (CDS). The floating diffusion capacitance was measured to be 20 fF, which has a noise contribution of 55 electrons. Spectral noise characteristics of the output amplifier are shown in Figure 7. This design utilizes buried channel transistors, resulting in lower flicker noise than an enhancement device of the same area. Noise is dominated by the small geometry input FET and produces a corner frequency of 2 MHz and a white noise component of 12 nV/√Hz. It is expected that the flicker noise is removed by an appropriate high pass filter in the signal processing chain. Referenced to the equivalent square pass noise bandwidth, the input referred buffer contribution is 30 electrons. The two uncorrelated noise components add in quadrature to yield the total read noise of 65 electrons. In this design, the buried channel full well is 110k electrons, resulting in a sensor dynamic range of 1500:1.

4.2 Image Smear

Smear plagues most image sensors to some degree and much effort has been devoted to reducing its effect. The interline transfer architecture is susceptible to two forms of

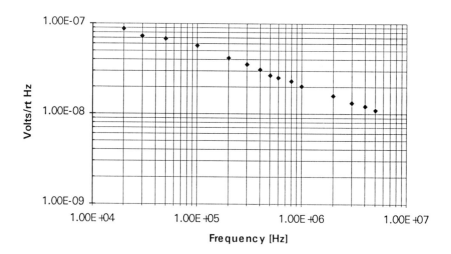

Figure 7 *Noise spectral density of the output buffer amplifier. The f^{-1} corner frequency is approximately 2 MHz, with a g_m noise of 12 nV/√Hz. Neglecting the f^{-1} component and referenced to a MHz bandwidth, this corresponds to noise equivalent electrons.*

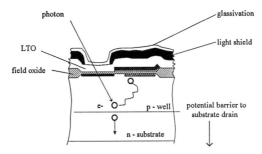

Figure 8 *Device cross-section illustrating the smear mechanism due to charge diffusion. Photo-electrons generated in the undepleted p-well are able to diffuse into the potential well of the adjacent vertical shift register. Minimizing the p-well thickness will preferentially direct the diffusion charge into the n- substrate drain.*

smear: (1) *diffusion smear* in which diffusion charge from the pixel mixes with the previous frame within the VCCD, and (2) *scattering smear* due to wave-guide effect of light through surface structures into the VCCD. These effects limit effectiveness of the anti-blooming and must be controlled to obtain good image quality over the desired exposure conditions.

Reticon has addressed the first of these mechanisms in the present design. The p-well depth of the VCCD was optimized to allow acceptable transfer efficiency, yet sufficiently shallow so as to minimize capture of lateral diffusion from the photosite (Figure 8). Diffusion into the region of the VCCD itself is proportional to the ratio of the p-well depth within the pixel diode region to the p-well depth of the VCCD. Minimizing the latter will force diffusion charge into the n-type substrate, essentially providing a small effective capture cross-section for the VCCD.

Smear from topside scattering arises from light piping and diffraction (Figure 9). This type of smear is mitigated by reducing the LTO layer thickness between the light shield and the underlying layers. Light piping decreases gradually with a thinner LTO thickness, while diffraction effects have a threshold value related to the wavelength of light and the index of the LTO layer. The LTO used in this process is nominally 1 μm thick. The

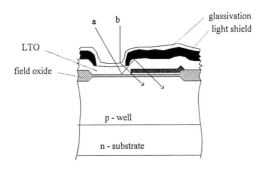

Figure 9 *Device cross-section illustrating the smear mechanism due to top layer effects. Ray (a) represents the wave guide effect and (b) illustrates diffraction.*

threshold thickness for blue light is ~1500 Å, indicating that this process is susceptible to light piping.

Measurement of the smear performance for this array was made by use of a pulsed light source, to time-tag the occurrence of illumination. To enhance accuracy of this measurement, the pulsed source was of sufficient intensity to over-saturate the device by a factor of 100. In Figure 10(a), the output of the array is shown in the over- illuminated condition. Smear signal will appear in the frame that precedes the frame containing the primary signal packet; in Figure 10(b) this frame shows no measurable signal. Limited by

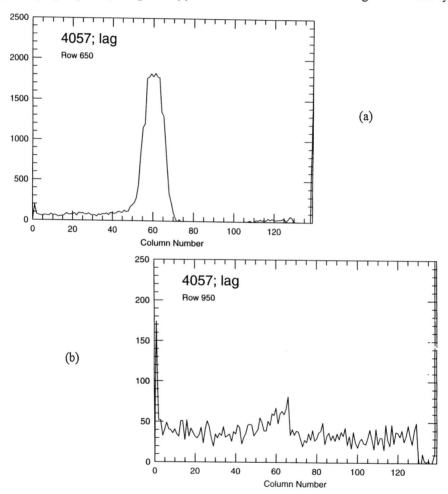

Figure 10 *Measurement of image smear in the 512(H) by 512(V) ILT Imager. (a) shows the video signal from one column over-illuminated by a factor of 100x. The light source is pulsed once every fourth frame. Image smear should be apparent in the video frame, which is time coincident with the pulse. This video signal is shown in (b); smear level is below the measurement limit and is below 0.01%, relative to a picture height of 10%.*

the SNR of the measurement, the indicated smear is less than 0.01%. This performance level is explained by the relatively large pixel pitch, which allows a significant light shield overlap of the vertical CCD, and the thin p-well of the vertical CCD.

4.3 Spectral Response

The pinned photodiode pixel design has a wide spectral response, extending from below 250 nm in the UV through NIR wavelengths of 800 nm. Long wave attenuation is due to use of the p-well design. Because of its depth relative to the absorption depth of photons in NIR wavelength band, the structure exhibits reduced photon-to-electron conversion efficiency within this spectral region. Measured response of the fabricated test array is shown (Figure 11) in a plot of absolute quantum efficiency versus wavelength.

In the current design, only 25% of the total pixel area is photosensitive. The remaining region is occupied by the non-active vertical CCD and channel isolation regions. Obviously, a lower active area results in decreased sensitivity; thus, Reticon will develop an on-chip microlens to regain most of the pixel aperture. We expect to achieve an effective fill-factor of 80% which will result in an increased QE, as illustrated in Figure 11.

6 SUMMARY

The authors have designed a second generation fast framing CCD imager in a 2/3" format (Table 1). The high performance design demonstrates further advancement in the use of

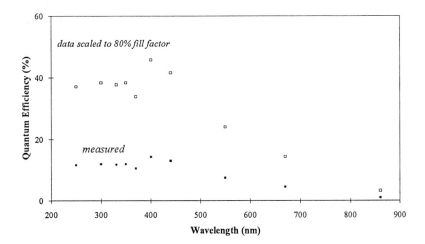

Figure 11 *Quantum efficiency measurement of the high speed 512(H) by 512(V) interline transfer CCD. Without a microlens, the QE peaks at 14%, due to the low fill-factor of an interline transfer design. Use of an integrated microlens will increase the effective fill-factor to ~80% as shown.*

Table 1 *Performance Characteristics of the Fast Framing Array*

Format	512 x 512
Pixel pitch	16 µm square
Image area diagonal	11.6 mm
Architecture	interline transfer
Vertical clocking	2 φ, 7V
Horizontal clocking	2 φ, 5V
Number of outputs	8
Frame rate	1,000 fps
Anti-blooming capacity	1000x
Dynamic range	65 dB
Readout noise	
amplifier	30e-
reset	55e-
quadrature sum	65e-
Spectral response range	200 nm - 800 nm
Dark Current	300 pA cm^{-2} (25°C)
Data rate per output	> 40 MHz
Output amplifier bandwidth	150 MHz

CCDs for the application.[2-4] Major design features of dynamic range, amplifier bandwidth and anti-blooming have been verified in operation and the architecture appears suitable for a wide variety of high speed imaging applications.

ACKNOWLEDGMENTS

The work described in this paper was supported in part by the Department of the Air Force under contract F088635-94-C-0002. Technical support was provided by Captain Patrick Harrington (USAF) of Wright Laboratories Armament Division, Eglin Air Force Base. The authors also acknowledge support and contributions Donald Snyder (USAF) also of Wright Laboratories.

References

1. D. Snyder and W.J. Rowe, SPIE Proc., 1990, **1346**.
2. T. Graeve and E. Dereniak, Vol. 1801, 1992.
3. R. Bredthauer, SPIE Proc., 1989, **1155**.
4. S.J. Strunk, *et al.*, SPIE Proc., 1993, Vol. 2002.
5. N. Teranishi, *et al.*, *IEEE Trans. of Electron Devices*, 1984, **ED-31**, No. 12.
6. P. Grey and R. Meyer, 'Analysis and Design of Analog Integrated Circuits', John Wiley and Sons, 1984.

A THREE SIDE BUTTABLE, 2048X4096 CCD IMAGE SENSOR

M. M. Blouke, T. Dosluoglu, and R. German
Scientific Imaging Technologies, Inc.
Beaverton, OR   97075-0569

S. T. Elliott
Jet Propulsion Laboratory
Pasadena, CA   91109

J. R. Janesick
PixelVision, Inc.
Huntington Beach, CA 92649

R. Reed
National Optical Astronomy Observatories
Tucson, AZ   85726-6732

R. Stover
UCO/Lick Observatory
Santa Cruz, CA 95064

1 INTRODUCTION

There are a number of applications which desire or even require a large focal plane: fluoroscopy, x-ray crystallography, astronomy, etc. Naturally, it is desirable to cover this focal plane with detecting material. This is especially important for astronomy where it costs many thousands of dollars/night to operate an instrument. Photographic film is one medium which can be manufactured in very large sizes. Amorphous silicon arrays in the 60x60 cm range and potentially larger are now being manufactured.

For CCDs to cover a large area there are several strategies. The most straight forward method would be to make a single device on a single very large wafer. This has some very hard limits today in the sense that the largest wafers commercially available are 400 mm (8") wafers. No one is currently fabricating scientific quality CCDs on these wafers. Several IC fabs that do make CCDs are utilizing 150 mm wafers. Consequently the largest chip that could be manufactured is limited to something on the order of 125 mm on the diagonal. This is a very large focal plane. If one reads the device in the slow scan mode such a chip will require a very lengthy time to readout.

An alternative is to pave the focal plane with a number of smaller chips. This has two advantages. The smaller chips are less expensive. Secondly, one achieves a natural multiplex advantage during the readout in that each chip is read out separately. The disadvantage is that there are non-imaging gaps between the chips, although with fiber optics these gaps can be minimized or eliminated.

The SI002A is a new CCD that is designed to fit into a mosaic of chips in order to assemble large focal plane arrays. In section 2, the layout and fabrication of the chip is described. Section 3 discusses measurements taken on two devices: a front-illuminated part and a back-illuminated part. Finally, the results are summarized in section 4.

2 ARCHITECTURE AND FABRICATION

2.1 Layout and Design

As indicated above the medical, nondestructive testing and astronomical communities all have need for high resolution, low noise, back-illuminated imagers. The SI002A was designed to fill that need.

The chip has 2048x4096, 15 μm square pixels arranged as two 2048x2048 imagers, one on top of the other. The architecture is basically a frame transfer configuration. An overflow drain is located at the top of the array to prevent any dark current generated in the field region from accumulating in the top row. Similar protection is afforded the edges of the device where single dummy columns on each side to again prevent dark current from the field region from accumulating in the first and last columns. Figure 1 presents a schematic diagram of the SI002A device.

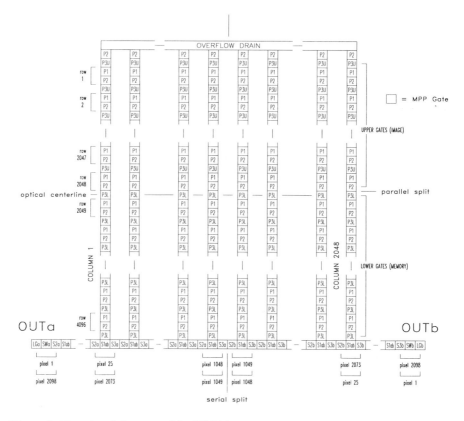

Figure 1 *Functional diagram of the SI002A*

Four parallel clocks are used to effect charge transfer: a common phase one and phase two clocks and each section has a separate phase three clock. Clocking both the parallel phase three clocks together allows full frame operation. The device is designed for MPP operation with the phase three (poly three gates) being the MPP phase. Frame transfer operation is effected by holding the phase 3 clocks low in the upper, imaging section while clocking the memory section. In this manner, signal charge in the imaging section is moved from phase one to phase two gates but remains confined to its respective pixel by the barrier phase three.

The serial register is located at the bottom and along one of the 2048 edges of the device. The serial register contains 25 overscan pixel at each end and is bent 90° away from the parallel section. The bend is necessary to provide room for the overscan pixels, and at the same time to conserve space so that the imaging array is as close to the edge of the chip as possible. The register is split in the middle so that the entire device can be read out either one or both of the amplifiers that terminate the serial register. The amplifiers are single stage MOSFET source followers based on a design that has proven to give between 2-4 electrons rms. read noise at 50 kpixel/s data rates. The serial register is designed with twice the well capacity of a parallel pixel and is terminated in a last gate and a summing well. The summing well is the same size and charge capacity as a normal serial gate.

The bond pads are all brought out on the end of the chip next to the serial register. The chip size is approximately 6.34x3.15 cm². Two such devices fit on a 100 mm wafer.

2.2 Device Fabrication

These devices are fabricated using the three level polysilicon gate technology with which all the standard devices at SITe are fabricated. The CCDs are built on 20-40 Ω-cm p-type silicon epitaxial wafers. In the fabrication process, the CCD channels are first defined by a p+ channel stop diffusion and field oxide. A buried channel is then formed by an n-type implant. Next a gate dielectric is grown, a layer of polysilicon deposited and the first level electrodes defined and etched. The exposed gate dielectric is etched and regrown and a second polysilicon layer deposited, patterned and etched to form the second level gates. This process is repeated a third time to produce the third and final level of gate electrodes. The output diodes and MOSFETs are formed by an n+ diffusion. An insulating layer of reflow glass is deposited, followed by metallization for the interconnections. This completes the processing for a front-illuminated device.

In order to fabricate a back-illuminated sensor, additional processing is required. The surface of the 525 μm thick silicon wafer containing the CCD structure is attached to a glass/ceramic substrate material. The resulting sandwich is turned over and all but 15 to 20 μm of the original wafer is removed. The device regions are photolithographically patterned and etched to isolate the individual chips. The back surface receives an enhancement treatment to reduce recombination of the photogenerated carriers near the exposed surface. An antireflection coating is applied. This may be either a coating designed for enhanced visible response or a coating which extends response to 200 nm in the ultraviolet. The support of the device using the glass/ceramic substrate has several advantages. First, the thinned device is completely supported and very rugged. Secondly, by mounting the device in this manner it is possible to use standard semiconductor

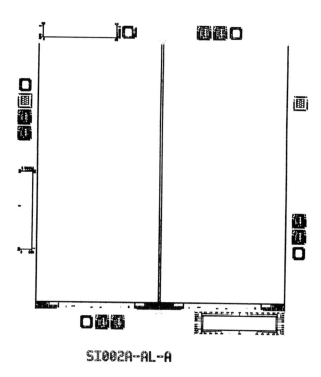

Figure 2 *Schematic of a wafer containing two SI002A devices and test structures*

processing techniques and equipment to process the devices through the thinning process. Since multiple devices are handled in parallel, this process will produce less expensive chips.

Processing options for the devices include MPP operation, minichannel implant and front- or back-illumination with visible or UV optimized AR coatings. Figure 2 presents a drawing of the layout of the devices on a wafer.

2.3 Packaging

The packages for these devices are based on a design developed by Lesser at the Steward Observatory, University of Arizona [1]. Figure 3 shows an overview of the package design with a mounted chip. The goals of the package are to minimize the spacing between chips when used in a mosaic. At the same time it is necessary to protect the edges of the chip. Although silicon is a very strong material, it is fairly easy to inadvertently nick the edge of a die. Since the active imaging area of these devices is so close to the edge of the die, some protection is desirable. Lastly, the package must be easily attached to some larger focal plane.

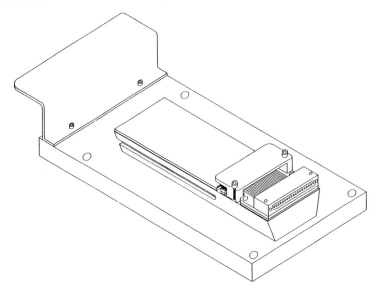

Figure 3 *Drawing of the packaged SI002A including handling substrate*

The chip is mounted on an Invar 36 chip carrier that is 50 μm wider than the chip on the long (4096) side. In addition, the carrier extends 25 μm beyond the top to the chip. The chip is soldered to the carrier with an Indium alloy. Assuming the chip is centered on the chip carrier, the separation between active imaging areas on two adjacent chips along the 4096 side will be less than 650 μm and between imaging areas on the short side, less than about 250 μm.

The lower end of the chip carrier extends beyond the soldered die. This shelf provides a place on which to mount a flexible circuit that connects the die to a 37 pin Nanonics connector. This connection is illustrated in Fig. 4. Note that the bonding pattern is symmetric. This means that the pinout for the front- and back-illuminated versions of the device are identical. It requires 32 pins to operate the device including 2 pins used for an on-chip temperature sensor. In addition, provision is made for mounting an optional AD590 chip on the flexible circuit to measure temperature. The chip ground is connected to the Invar 36 chip carrier for both front- and back-illuminated parts.

For handling purposes, the chip carrier is mounted on a large Al substrate with a handle. A 37 conductor ribbon cable with a shorting plug is connected to the Nanonics connector. Connection to the device for testing purposes is made through this removable cable.

3 TEST DATA

Table 1 presents a summary of the performance expected from these devices. A number of front- and back-illuminated devices have been fabricated. Several have been thoroughly been characterized. We report results from several devices here.

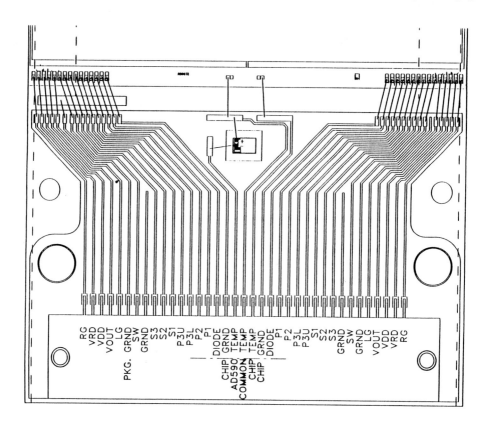

Figure 4 *Detailed view of the pinout of the 2kx4k CCD*

Table 1 *Expected CCD performance*

Parameter	Value
CTE	
Parallel	>0.99999
Serial	>0.99999
Noise	2-4 e rms.
Conversion gain	~1.8 µV/e
Full Well	~100 ke
Dark current	
MPP	< 50 pA/cm²
non-MPP	<250
Quantum efficiency	see Fig. 8
Flatness	± 5 µm

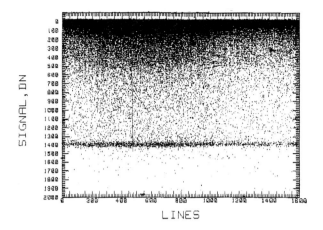

Figure 5 *Vertical stacked line trace using Fe^{55} x-rays CTE is >0.99999 in both directions*

Figure 5 presents Fe^{55} x-ray data for the vertical charge transfer direction for a front-illuminated device. As the data indicate, the CTE is greater than 0.99999. Similar measurements in the horizontal direction yielded an equally high transfer efficiency as did measurements on the back-illuminated part. Measurements using the extended pixel edge response technique indicate parallel CTEs in excess of 0.999998 at the 10000 e signal level.

Full well measurements on the front illuminated part indicated an optimum well capacity of 90000 electrons. Measurements on other parts have given values between 70 and 100 ke.

The noise for several devices has been measured. For a front illuminated part, the read noise of one amplifier was 3.2 e- rms. for a 4 µsec clamp-to-sample time (the other amplifier was nonfunctional). For a back-illuminated part the noise was 2.9 e- using an 8 µsec clamp-to-sample time on one amplifier and 4.7 e- on the other amplifier. The conversion gain for this amplifiers was 1.84 µV/e. In all cases the data rate was 50 kpixel/s. Measurements on other devices has yielded values from 3 to 9 e.

Dark current is well behaved. For the back-illuminated part, the dark current was measured to be 27 pA/cm^2 in the MPP mode. These measurements are shown in Fig. 6. Measurements on other MPP parts operated in the non-MPP mode yield values as low as 3-5 e/pixel/hr at -80°C. This translated to approximately 150 pA/cm^2 at 293 K.

Figure 7 presents quantum efficiency data for a back-illuminated part. These data were taken at -90°C. The device had SITe's standard visible AR coating. The peak quantum efficiency was about 85%.

The bow of a number of thinned devices has been evaluated. The bow on selected parts has been measured from between 80 to 180 µm and is largely spherical in nature. It is important to note that no effort has been expended to flattening these devices. It is expected, however, that flatness ±5 µm can be achieved with this package and with modified packaging techniques. Efforts are currently in progress to achieve these results.

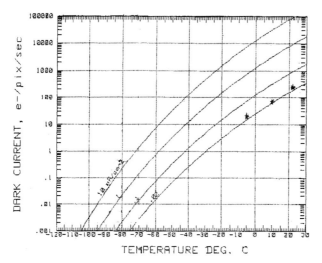

Figure 6 *Dark current as a function of temperature for a back-illuminated SI002A device*

4 SUMMARY

The SI002A is a large area, scientific quality, CCD image sensor intended for use in the medical, nondestructive testing, and astronomical communities. The device has 2048x4096 pixels in the serial and parallel directions, respectively. The chip and package are designed to be three side buttable with leads or electrical connection to the chip occupying only one side of the die. A number of front- and back-illuminated parts have been fabricated and tested. In general, the device is well behaved with low noise, high CTE, high quantum efficiency and low dark current.

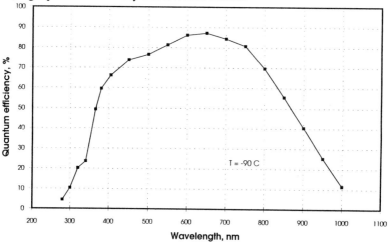

Figure 7 *Quantum efficiency as a function of wavelength for a back-illuminated part operating at -90 °C*

ACKNOWLEDGMENTS

The authors would especially like to thank P. Marriott and T. Yamaguchi of SITe for technical assistance in the fabrication of these devices. They would also like to thank Dr. Michael Lesser of the University of Arizona for many useful discussions regarding the packaging of these devices. The authors would also like to thank T. Wolfe, and Dr. C. Claver of NOAO and M. Wei of Lick Observatory for their assistance in the evaluation of certain devices.

References

1. Dr. M. Lesser, private communication.

BACK ILLUMINATED CCD PROCESSING AT STEWARD OBSERVATORY

M. P. Lesser

Steward Observatory
University of Arizona
Tucson, Arizona 85721

1 INTRODUCTION

Charge-Coupled Devices (CCDs) are the primary imaging detectors at nearly every major astronomical observatory as well as for many other scientific applications. Commercially available CCDs, however, leave significant room for improvement before they can be considered as ideal detectors. In particular, front illuminated devices have low quantum efficiency (QE), especially in the blue and UV. Even the largest CCDs are relatively small compared to typical telescope focal planes. And most are not packaged in a manner conducive for many instruments such as spectrographs and very large area imaging cameras.

The Steward Observatory CCD Laboratory (CCD Lab) at the University of Arizona has been developing the technology to produce back illuminated CCDs from commercially manufactured devices over the past six years. This work has led to the development or improvement of several related processes, such as antireflection (AR) coatings, backside charging techniques, and device packages and packaging techniques. In this paper, we present a review of these developments, with emphasis on our recent work to develop buttable focal plane mosaics using foundry-produced CCDs.

2 BONDING AND THINNING

Our back illuminated processing starts with CCD wafers which are commercially lapped to remove about half of the silicon substrate material. This step allows better cosmetics in the final devices because less etching is required. The final thickness is chosen such that the wafers are strong enough for handling. After lapping, the CCDs are diced and only those die chosen for back illuminated operation are further processed.

After dicing, each CCD is flip chip bonded onto a custom silicon substrate manufactured to provide electrical I/O from the bonding pads on the CCD to wire bonding pads which lead to the package in which the CCD is to be mounted.

We have modified a Research Devices flip chip bonder to allow full 4" wafer bonding in order to process the large CCD manufactured today (Figure 1). As CCDs become larger, the chucks on this machine can be further enlarged to bond even larger die. We use gold and/or indium bumps in the flip chip bonding process. The bump sizes are set by the

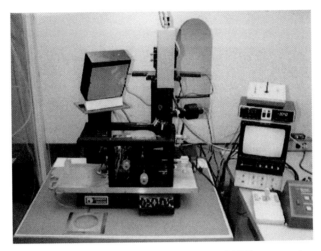

Figure 1 *Modified flip chip bonder for back illuminated CCD processing*

bonding pad areas and the die size. Large devices need higher bumps in order to overcome the intrinsic warpage of the die. We use vacuum chucks to flatten both the substrate and die, although the custom substrates are so flat the major contribution to device non-flatness is the die itself. By using the flip chip bonding process we are able to achieve a surface flatness of $\pm$ 5 μm with a 2048^2 15 μm pixel CCD.

After flip chip bonding, an epoxy is used between the CCD and substrate to provide mechanical support after the device is thinned. This epoxy has been chosen to remain fairly flexible at cryogenic temperatures and to have a low viscosity for the underflow process. After curing, the CCD can be thinned.

Thinning occurs in a selective HF-based acid using a linear agitation thinning machine developed by our Lab. The etch used slows greatly when the epitaxial layer of the die is reached. This is typically 20 μm thick for most CCDs. Unfortunately, the CCDs are depleted to less than 10 μm, which results in a nearly 10 μm field-free region if thinning is stopped just after reaching the epitaxial layer. This field free region allows charge spreading to occur in which the photogenerated electrons can diffuse to adjacent pixels with a resultant loss in resolution or Modulation Transfer Function (MTF) degradation. Further thinning into the epitaxial layer is desirable for the highest MTF, though red QE will be lower and interference fringing will be worse. We have recently added an additional (optional) thinning procedure which allows fine control of the final thickness of the CCD to tailor MTF for specific applications.

3 PACKAGING

The standard commercial CCD packages, which are usually used for cryogenically cooled CCDs, are made of stamped Kovar with electrical pins glass-sealed to the package (Figure 2). Wire bonds are made from the CCD bonding pads to these pins. We have found such packages to be unsuitable for many CCDs because they do not easily allow for edge butting without special machining and because the Kovar is warped in the forming process

Figure 2   *Commercially available Kovar packages used for some CCD Laboratory CCDs*

and does not allow a flat device to be produced. We have measured commercial packages to be warped by more than 200 µm peak-valley using a 10.6 µm laser interferometer. We have, therefore, developed several alternative packages which we now routinely use for scientific CCDs.

Our custom package concept is to use an Invar base for mechanical support, a printed circuit for I/O fanout, and a micro-connector for the external electrical connection (Figure 3). This system allows very general packages to constructed, buttable when necessary, with the required tolerances.

We use Invar-36 as the stock material from which to fabricate our packages. This iron/nickel alloy is readily machinable and available. It is well received by machine shops and generally not considered an "esoteric" material. The packages are machined and surface ground with a flatness and parallelism specification of 0.0002". The buttable edges of the package are generally undersized relative to the imager by 0.010" on a side. Typically, four mounting holes are provided, two tapped holes on the underside of the buttable edge and two oversized clearance holes opposite the tapped holes on the non-butting edge. The advantage to this scheme is appreciated during mosaic assembly. With two clearance holes in the package, one can align the CCD to the desired location on the baseplate under a microscope. This location is fixed by threading in the two screws from above. After all devices have been secured by two screws, the underside screws are inserted through oversized clearance holes in the baseplate and threaded into the package. This firmly locks all the packages on the baseplate with four screws.

The CCD die is attached to the Invar package with an thermally conductive epoxy or thermoplastic. The thermoplastic is set by heating the package, typically to about 150°C. The maximum operating temperature of the final CCD package should never exceed (or meet) this temperature to avoid movement of the CCD relative to the package. The thermoplastic has many advantages over dispensed epoxy, including increased ease in packaging, cleanliness, and improved thermal contact between die and package.

Figure 3 *Custom machined Invar/printed circuit packages used for many buttable CCDs processed in the CCD Laboratory*

We use Nanonics high density cryogenic connectors for electrical connection between the PCB and an external wire harness (Figure 4). The connectors accept mating plugs which are purchased with an attached cable. Electrical connection to each CCD, therefore, consists of plugging in the connector. We have tested these connectors in LN_2 and found them to be very reliable. Our test dewar has used such a plug and wire hardness for over 500 cycles with no noticeable changes. The cable we presently use has 30-gauge stranded copper wires, although higher thermal resistance wires can be special ordered. We also have had custom shorting plugs manufactured for shipping and storage purposes. We have used both 37- and 15-pin connectors on 0.025" centers.

Using these techniques, we have made several different edge-buttable CCD packages for Loral/Fairchild CCDs. These include the 2048^2 3-edge buttable, the 2048^2 2-edge buttable, and the 2kx4k 3-edge buttable. We show in Figure 5 some of these packages with CCDs mounted.

4 BACKSIDE CHARGING

After a CCD is thinned, a silicon oxide film grows naturally on the backside due to contact with air. Unlike the finely controlled oxides used in the semiconductor industry for integrated circuit manufacture, this native oxide contains impurities which result in a net positive charge in the oxide. In addition to this positively charged oxide, there are dangling molecular bonds at the $Si-SiO_2$ interface due to the abrupt termination of the silicon crystal lattice. This creates positive surface charge as well. This positive charge creates a potential well at the backside which has a significant effect on free electrons within several thousand Ångstroms of the $Si-SiO_2$ interface. Any photon absorbed within this region produces a photoelectron that has only a small chance of migrating to the frontside collection wells. The photoelectron will most likely be trapped in the backside

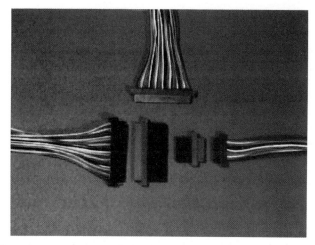

Figure 4 *Nanonics connectors and shorting plugs used for electrical I/O with custom packages*

potential long enough to recombine. Photons with a wavelength of $\lambda < 5000$ Å are significantly affected, since the absorption depth of silicon decreases rapidly with decreasing wavelength. Therefore a freshly thinned CCD's initial QE is much poorer than even a front illuminated device.

One method of eliminating this backside potential well is to counteract the positive charges in the oxide by applying a layer of negative charge to the backside of the CCD. For this process to be effective, it is necessary to tie-up most of the dangling bonds in order to reduce the backside well and restore reasonable QE. This passivation process is accomplished by growing a thin oxide on the CCD backside.

Figure 5 *Several processed CCDs on custom packages*

Because of the indium and epoxy in our flip chip bonding process, our CCDs cannot be heated to a temperature of more than about 150°C. We are therefore constrained to use a very low temperature oxidation technique to passivate the back surface. We have developed several modifications to the traditional wet oxide growth process which allow uniform and reproducible oxidation at very low temperature. Using this process, we are able to grow an oxide of sufficient thickness for backside charging in a few hours. For long term stability, however, a thicker oxide is required, and we typically grow the oxide for 24 hours. This oxidation process allows backside charging to yield the theoretical maximum QE at all wavelengths in the 200 - 1100 nm region.

After proper backside oxidation, the CCD can be backside charged using either a UV light flood or by deposition of a thin platinum film. The uniformity of the backside charge is mainly determined by the uniform quality of the backside oxide. We typically achieve uniformity of less than 5% spatial QE variation.

The UV flooding technique theoretically yields the highest QE possible with a CCD since there is no backside absorption layer of back surface damage. The process is described in detail elsewhere[1,2,3]. It basically consists of shining a UV lamp on the CCD prior to cooling in a vacuum dewar. The photogenerated electrons trap oxygen on the back surface, which creates a negative surface charge layer. This charge is maintained indefinitely when the CCD is cool, and drives photogenerated electrons to the frontside for detection during operation. The drawbacks of the process are the requirements for a UV lamp and UV transparent optics, and access to a vacuum pump and dry O_2 source. The extremely high QE obtained, however, has made this a standard process in the astronomical community.

Because of the drawbacks mentioned, we have continued work begun by Janesick at JPL[4] to utilize a platinum film to charge the CCD backside surface. This process uses platinum's higher work function (than silicon) to cause electrons to tunnel through the backside oxide and charge the back surface much as is done with UV flooding but without the UV lamp. We have in fact found this process relies significantly on the

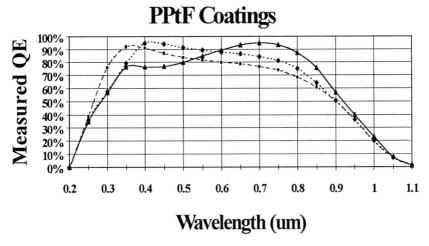

Figure 6 *Measured QE of several passivated platinum film coated devices optimized for different wavelengths*

Steward Observatory CCDs

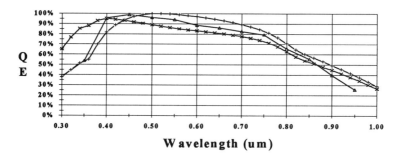

W avelength (um)

Figure 7 *Measured QE of several UV flooded devices with different AR coatings*

reactivity of the Pt rather than just the work function difference. We have developed a multi-layer passivation and coating process we call a Passivated Platinum Film (PPtF)which has very stable and uniform charging properties. While the PPtF can still be discharged with contaminants in a vacuum dewar, it can be simply recharged by the introduction of O_2 for a short time or even by simple vacuum pumping to desorb contaminants. The PPtF QE is extremely high, as shown in Figure 6, and for many applications is preferred to the UV flooded coatings because of its simpler operational requirements.

5 ANTIREFLECTION COATINGS

The application of an antireflection (AR) coating to a back illuminated CCD can increase the QE of the device to nearly 100% in the visible spectral region. We typically use hafnium oxide (HfO_2) as the coating material because it is UV transparent and compatible with UV flooding and platinum gates. A single layer coating can be optimized to provide peak QE at any wavelength in the 200-800 nm region. We have also developed a 2-layer HfO_2 and MgF_2 AR coating which yields a very high average QE over a broad spectral region. This coating has been applied to most of our edge-buttable CCDs. We show the measured QE of several AR-coated devices in Figure 7. These curves show a good representation of the various coatings which can be applied to different CCDs, each optimized for the observational requirements required of the detector.

References

1. M. P. Lesser, A. Bauer, L. Ulrickson, and D. Ouellette, Proc. SPIE, 1992, **1656**.
2. M. P. Lesser, "Backside Coatings for Back Illuminated CCDs", Proc. SPIE, 1993, **1900**, 219.
3. M. P. Lesser, M. P., Proc. SPIE, 1994, **2198**, 782.
4. J. Janesick, T. Elliot, T. Daud, and D. Campbell, *Optical Engineering*, 1987, **26**, 852.

APPLICATION SPECIFIC IMAGER DESIGN AND FABRICATION AT SARNOFF

John R. Tower

David Sarnoff Research Center
CN 5300
Princeton, NJ 08540-5300

1 INTRODUCTION

David Sarnoff Research Center has developed a number of integrated circuit fabrication technologies to support our custom imager and camera development business. These include a UV-visible, spectroscopic CMOS/CCD process, both CMOS and CCD infrared PtSi focal plane processes, and a variety of thinned, backside illuminated processes. These processes have been utilized to produce imagers for commercial analytical spectrometer instruments, military thermal imaging systems, and multiple-use high frame rate visible camera systems, respectively.

The MOS-based solid state imager development at Sarnoff began in the 1960s. The CCD development effort began in 1971. Sarnoff, as RCA Laboratories and then as a subsidiary of SRI International, has had a continuous effort in MOS and CCD technology for more than 25 years. Sarnoff researchers hold more than 100 patents in these fields, including the CCD floating diffusion output and the notch (trench) to improve CCD transfer efficiency. Today's efforts support a variety of imager technologies which permit solid-state imaging from the UV to the medium wave infrared, and pixel output rates exceeding 3×10^8 pixels per second. This paper provides an overview of our current activities.

2 SPECTROSCOPIC CCD TECHNOLOGY

Sarnoff has developed a segmented CCD tailored specifically for spectroscopic applications. The architecture of the device was developed in collaboration with Perkin Elmer.[1] The process technology is based upon a melding of RCA's CMOS and buried channel CCD technologies. The double level polysilicon, two level metal process incorporates N-well CMOS logic with high performance buried N-channel CCD registers and a proprietary virtual gate detector technology. The Perkin-Elmer spectroscopic CCD is shown in Figure 1 mounted in its ceramic carrier and flex circuit.

The spectroscopic CCD is employed in the Perkin-Elmer Optima 3000 analytical instrument. Two devices are employed for each instrument. One is dedicated to detecting the UV emission lines, while the other is dedicated to detecting the visible emission lines. Sarnoff has supplied hundreds of devices to Perkin-Elmer to satisfy the high demand for the Optima 3000.

The spectroscopic CCD consists of 224 linear arrays precisely placed to image the appropriate emission lines. The linear arrays have varying numbers of virtual gate detectors of varying detector heights. All of the arrays are readout through a single output port. The CMOS logic receives array addressing information via a parallel digital address port and appropriately connects the correct readout clocking pulses to the selected array, provides deselect-integration control biases to all of the other arrays, and connects the floating diffusion output of the selected array to the single output port.

The CCD design employs the second level metal as an optical shield. The serial CCD registers are two phase, double level polysilicon designs with an implanted barrier under poly 2. This implanted barrier reduces the number of required bias levels and simplifies clocking. The thin gate dielectric (350 Å SiO_2) and CMOS transmission gates for clock distribution permits use of 5 - 6V amplitude clocks.

The virtual gate detector technology provides excellent UV-visible quantum efficiency. Measured quantum efficiency is shown in Figure 2. For reference, the typical quantum efficiency of a photomultiplier tube (PMT) is also shown in Figure 2. The high quantum efficiency combined with a noise floor < 15 e RMS (O C temperature) provides excellent sensitivity.

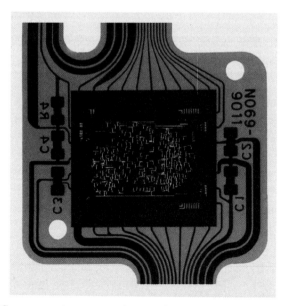

Figure 1 *Spectroscopic segmented array CCD in production configuration*

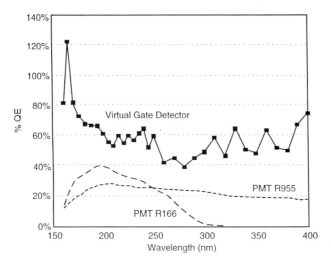

UV Spectrum

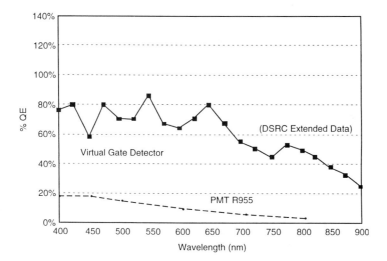

Visible Spectrum

Figure 2 *Quantum efficiency of spectroscopic CCD compared with PMT[1]*

3 PtSi FOCAL PLANE ARRAY TECHNOLOGY

Both CCD and CMOS-based Schottky barrier PtSi focal planes are produced by Sarnoff. The Schottky barrier infrared focal plane work at Sarnoff began under Air Force funding in 1974. The CCD process is presently utilized to produce a 320 (H) x 244 (V) interline transfer focal plane array with 40 micron x 40 micron pixels.[2] The device is processed with a cryogenic-compatible CCD process employing two levels of polysilicon and one level of metal. The optical fill-factor is 43%. The backside illuminated focal plane provides a noise equivalent delta temperature, NEDT < 0.05 C (30 Hz, f/1.8 optics). The pixel layout for the 320 x 244 focal plane is shown in Figure 3.

Since the silicon bandgap precludes transmission through the silicon below roughly 1 micron, the backside illuminated CCD is responsive from roughly 1 - 5.5 microns. Devices designed for thermal response in the 3 - 5 micron band have an optical cavity with reflector above the PtSi detector to optimize the performance in this band. An interesting variation on the standard 320 x 244 has been developed. Sarnoff produces a very wide-band focal plane array by topside illuminating a modified version of the 320 x 244 device. In this design the reflector is removed to permit radiation to reach the PtSi detector, and the interline transfer CCD register is optically shielded. This produces a focal plane which is responsive from the UV to the medium wave infrared. Figure 4 shows imagery produced by a topside illuminated 320 x 244 device in three different wavelength bands: the visible/near infrared, the shortwave infrared, and the mediumwave infrared. Typical quantum efficiencies in this mode of operation are 40% QE at 0.4 micron, 4% QE at 1.0 micron, 2% QE at 1.65 micron, and 0.3% QE at 3.4 micron.

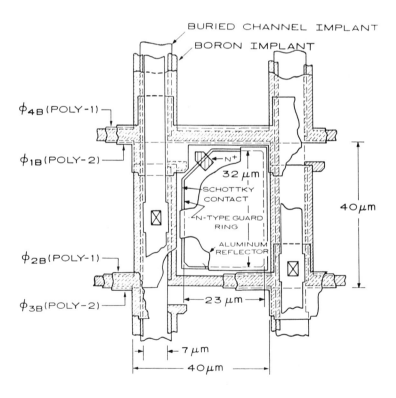

Figure 3   *Layout of detector array for 320 x 244 PtSi CCD focal plane array*

 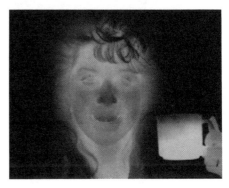

Figure 4 *Imagery from 320 x 244 visible/SWIR/MWIR PtSi focal plane array*

The CMOS PtSi process is utilized to produce a 640 (H) x 480 (V) subframing focal plane array with 24 micron x 24 micron pixels.[3,4] The device is shown in Figure 5. This device is produced on our 5X stepper with 1.2 micron design rules. The single level polysilicon, double level metal CMOS imager is capable of many operating modes including progressive scan or interlaced readout, integration time control, and subframing imaging. These operating modes are portrayed in Figure 6.

4 VISIBLE TECHNOLOGY

The standard visible CCD process technology at Sarnoff is a thinned, backside illuminated, triple polysilicon technology. Its heritage is the RCA process developed in support of surveillance applications, scientific CCDs, and broadcast quality CCDs. Sarnoff participated in the development of the process technology in the 1970s and early 1980s. RCA Laboratories (Sarnoff) was recognized with RCA Lancaster and RCA Broadcast by an Emmy Award for the world's first broadcast quality CCD camera in 1985. In the late 1980s, Sarnoff essentially moved the Lancaster production process to Sarnoff, and converted it from 3" to 4" wafers.

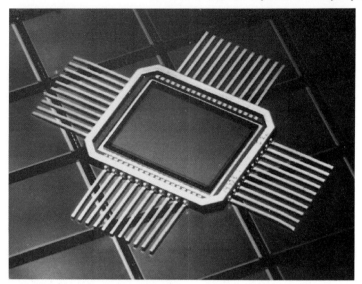

Figure 5   *640 x 480 PtSi subframing IRFPA*

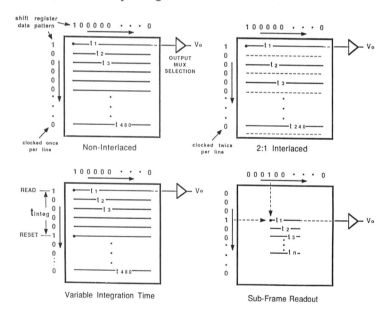

Figure 6   *Operational flexibility of 640 x 480 PtSi CMOS IRFPA*

The process employs whole wafer thinning, backside implantation, and a furnace anneal to activate the implant. This process establishes a stable backsurface electric field which provides high quantum efficiency, high uniformity, and a very high responsivity uniformity. After thinning, the wafers are laminated to a supporting glass disk.

Devices with and without blooming drains are produced. A cross-section of a device with buried blooming drains is shown in Figure 7. This blooming drain structure maintains a 100% optical fill-factor and provides >10,000X overload capability (vertical transfer smear affects not included). Typical quantum efficiency for a device with and without blooming drains is shown in Figure 8. These devices do not have AR coatings.

4.1 High Frame Rate Visible CCDs

Under Army Research Laboratory funding, a family of high frame rate, backside illuminated CCDs have been developed.[5,6] These devices have multiple output ports to permit high frame rate operation. The 512 x 512 device has 16 output ports and has been operated at rates up to 800 frames per second. The 1024 x 1024 device has 32 output ports and operates at rates up to 300 frames per second. These devices have on-chip correlated double sampling (CDS) and buried blooming drains for overload control. The architecture of the 512 x 512 device is shown in Figure 9.

4.2 Radiation Hardened Visible CCDs

Working with Rockwell, a radiation hardened visible CCD technology is being developed.[7] A 512 x 512 device with four output ports has been fabricated. The backside illuminated CCD employs a notch (trench) implant for proton displacement hardening and a thin oxide for total dose hardening. Devices with a notch and standard dielectric thickness have been tested and proven to provide improved proton displacement damage hardness. Devices with a notch and thin dielectric have been functionally tested and await radiation testing.

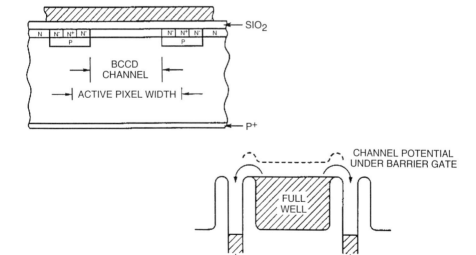

Figure 7   *Cross-section and potential profile for a backside illuminated device with blooming drains*

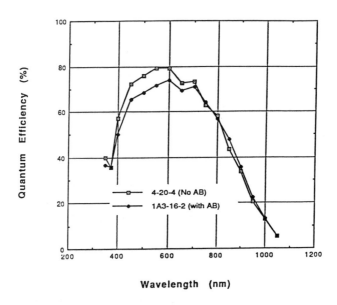

Figure 8   *Quantum efficiency for a backside illuminated device with blooming drains and a device without blooming drains*

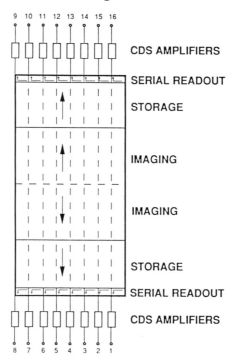

Figure 9   *Architecture of 512 x 512 high frame rate CCD*

4.3 Deep UV Area Array CCDs

Sarnoff has investigated two approaches to deep UV (<200 nm) area array CCDs. The first approach is to treat our conventional backside illuminated CCDs with a proprietary passivation. This passivation combined with our furnace activated, implanted backsurface treatment gives enhanced response in the UV compared to conventional backside illuminated CCDs.

The second approach is to implement a buried P-channel CCD.[8] This structure is identical to the thinned, backside illuminated buried N-channel CCD, except that the polarity is reversed in all respects. This approach allows the natural charging of the silicon backsurface to enhance the band-bending (electric field) at the back surface, rather than degrading the electric field causing trapping and recombination of carriers, as is the case in buried N-channel CCDs. Figure 10 shows the quantum efficiency achieved to date with our P-channel CCDs.

5 CUSTOM PROCESS DEVELOPMENT

As evidenced by the preceding discussion, Sarnoff supports a variety of process technologies. Sarnoff also develops custom processes for specific customer applications. An excellent example of this custom process development capability is a complex, high density process developed for very high speed imaging applications.[9]

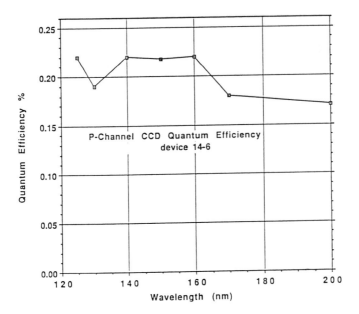

Figure 10 *Quantum efficiency of P-channel CCD*

For this high speed imaging application, a four level polysilicon, three level metal process was developed. The process also incorporates a high-transfer-speed virtual gate detector. The process supports 1.5 micron design rules running on our 5X stepper. The buried channel CCD process has been optimized to maximize the charge handling capacity and the fringing fields of the very small BCCD storage elements (1.5 micron x 3 micron storage gates). A Scanning Electron Micrograph (SEM) showing the process topography is shown in Figure 11.

6 SUMMARY

Sarnoff, formerly RCA Laboratories, has a thirty year history of innovation in solid state imaging. Today, Sarnoff continues to provide innovative application specific imaging solutions utilizing a variety of process technologies. A summary of Sarnoff's process technologies for imaging applications is given in Table 1.

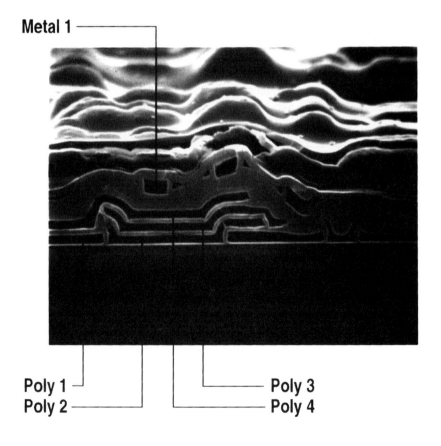

Figure 11 *SEM of four level polysilicon, three level metal very high speed imager technology*

Table 1 Summary of Sarnoff's Process Technologies for Imaging Applications

Process	Layers Polysilicon	Layers Metal	Pixel Technology
Spectroscopic CMOS/CCD	2	2	Virtual Gate Detector
Infrared CCD	2	1	PtSi Detector
Infrared CMOS	1	2	PtSi Detector
Visible CCD	3	1	Thin-Backside CCD
High Density CCD	4	3	Virtual Gate Detector

References

1. T. W. Barnard, M. I. Crockett, J. C. Ivaldi, P. L. Lundberg, D. A. Yates, P. A. Levine, and D. J. Sauer, *Analytical Chemistry*, 1993, **65**, 1231.

2. T. S. Villani, W. F. Kosonocky, F. V. Shallcross, J. V. Groppe, G. M. Meray, J. J. O'Neill III, and B. J. Esposito, "Construction and Performance of a 320 x 244-Element IR-CCD Imager with PtSi SBDs", *SPIE Proc.*, 1989, **1107-01**, 9.

3. T. S. Villani, B. J. Esposito, T. J. Pletcher, D. J. Sauer, P. A. Levine, F. V. Shallcross, G. M. Meray, and J. R. Tower, "Performance of Generation III 640 x 480 PtSi MOS Array", *SPIE Proc.*, 1994, **2225**, 2.

4. J. Ambrose, B. King, J. Tower, G. Hughes, P. Levine, T. Villani, B. Esposito, T. Davis, K. O'Mara, W. Sjursen, N. McCaffrey, and F. Pantuso, "High Frame Rate Infrared and Visible Cameras for Test Range Instrumentation", *SPIE Proc.*, 1995, **2552-37**, 364.

5. D. J. Sauer, F-L. Hsueh, F. V. Shallcross, G. M. Meray, P. A. Levine, G. W. Hughes, and J. Pellegrino, "High Fill-Factor CCD Imager with High Frame-Rate Readout", *SPIE Proc.*, 1990, **1291**, 174.

6. P. A. Levine, D. J. Sauer, F-L. Hsueh, F. V. Shallcross, G. C. Taylor, G. M. Meray, and J. R. Tower, "Multi-Port Backside Illuminated CCD Imagers for Moderate to High Frame Rate Camera Applications", *SPIE Proc.*, 1994, **2172**, 100.

7. P. A. Bates, P. A. Levine, D. J. Sauer, F-L. Hsueh, F. V. Shallcross, R. K. Smeltzer, G. M. Meray, G. C. Taylor, and J. R. Tower, "Radiation Hardened, Backside Illuminated 512 x 512 Charge Coupled Device", *SPIE Proc.*, 1995, **2415**, 43.

8. P. A. Levine, G. C. Taylor, F. V. Shallcross, J. R. Tower, W. Lawler, L. Harrison, D. Socker, and M. Marchywka, "Specialized CCDs for High Frame-Rate Visible Imaging and UV Imaging Applications", *SPIE Proc.*, 1993, **1952**, 48.

9. W. Kosonocky, G. Yang, C. Ye, R. Kabra, J. Lowrance, V. Mastrocolla, D. Long, F. Shallcross, and V. Patel, "360 x 360-Element Very High Burst-Frame Rate Image Sensor", IEEE International Solid State Circuits Conference, February 1996.

STATUS OF LARGE AREA CCD IMAGERS AT LORAL

Richard A. Bredthauer and Robert J. Potter

Loral Fairchild Imaging Sensors
14251A Chambers Rd.
Tustin, CA 92680

1 INTRODUCTION

This paper describes recent progress made for very large area scientific imaging arrays at Loral. CCD imaging arrays from 3cm x 3cm (2048x2048 pixels) to 8cm x 8cm (9k x 9k pixels), are discussed. These devices show exciting promise in a variety of medical, astronomical, and high resolution reconnaissance imaging applications. The digital imager evolution has culminated in the successful design, fabrication and demonstration of a fully functional 9216x9216 pixel array. With a total of 84.9 million pixels, this CCD is the world's highest resolution full frame imager.

Arrays as large as 3 centimeters on a side (2048x2048 pixels) were first demonstrated in the late 80's[1]. Those initial imagers were designed with a pixel size of 15μm. sq. specifically to maximize the fit of four devices per 4"(100mm) wafer. CCDs have one of the largest active gate areas of any semiconductor device. Creating a design which is tolerant to predominant process defects is the key to yields for devices of this size. As a result, process yields are reaching a point where full wafer imagers are not only possible, but are becoming economically feasible. The evolution of process technology continues to drive down the price of devices while improving the quality. In the next three sections we describe how large area scientific imagers have become the dominant digital imaging technology.

2 HIGH RESOLUTION ASTRONOMICAL IMAGING

Astronomers, were the first to recognize the potential of the CCD for high quality scientific imaging. In comparison to photographic film and SEC vidicon tubes originally used, CCDs offer several benefits to the astronomer. CCDs provide long term stability and a direct digital output allowing for very rapid data reduction. With a sensitivity of 100 times faster than film, it is clear that more data can be produced in a shorter period of time. Within a few years the CCD has become the sensor of choice at all major observatories (with the possible exceptions of Schmidt telescopes which utilize very large photographic plates). The most dramatic demonstration of the CCDs performance is on the Hubble Space Telescope. Eight 800x800 15 micron Loral CCDs are flying on the telescope as part of the WFPCII camera. They provide near real-time imagery of outstanding resolution.

Figure 1 *Hubble Space Telescope*

Over the past five years Loral has produced a wide variety of custom scientific imagers for astronomical applications. The majority incorporate 15 micron square pixels. These include 4000x200, 3000x1536, 3072x1024, 2688x512, 2560x256, and 1200x800 configurations for spectroscopic applications. There are variations of a 2048x2048 imager. Versions incorporate provisions for two adjacent side abutting or three side abutting[2,3,4].

Many of proposed instruments will employ 2 x 2, 3 x 3, or even larger mosaics of medium-to-large format CCDs, which present technical challenges for both the focal plane and dewar designs. A trick in beating the yield problem and obtaining ultra-large CCDs is to mosaic them. The CCDs are designed so that non-shorted devices can be diced, butted, and mosaiced on two or three edges that can be organized into a 4096 x 4096 pixel, or larger format[5]. Packaging costs are high when using this approach since very tight tolerances are required for butting the chips and keeping the seam regions between each device to a minimum (a couple pixels can be achieved). For similar packaging efforts it may be more cost effective to fabricate a full wafer CCD rather than resorting to mosaicing.

As the process technology matures and the wafer size increases the mosaic consists of larger and larger individual CCD devices. The present *de facto* astronomy standard device is a 2048x4096 15μm three side abuttable array. This scientific imager is manufactured by Loral, Orbit[6] and SITe. This is presently the optimum cost/yield tradeoff for practical production. We are now beginning to see reasonable yields of 4096x4096 pixel imagers on 100mm and 125mm wafers.

The 9k imager offers a tremendous advantage for the large field (12"x12") high resolution Schmidt telescopes. A staggered array of four devices would cover a 5 degree field of view with a sampling of 0.55 arc seconds per pixel. Data handling becomes the biggest issue. A single frame from one 9k imager produces a 168 megabyte image, the

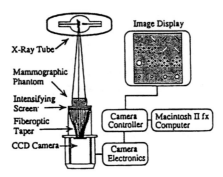

Figure 2 *Schematic of fiber optic CCD X-ray system[10]*

equivalent of 132 Hubble images. One hour (120 four imager frames) would produce more than 80 gigabytes of data[7].

3 LARGE AREA MEDICAL IMAGERS

Medical X-ray imaging is another area in which film is rapidly being replaced by large area digital CCD imaging devices. The requirements are similar to astronomy, requiring large dynamic range, high quantum efficiency, low dark current, and stable solid state operation. Medical applications include biological X-ray microscopes, medical X-ray cameras (e.g., spot mammography) and dental X-ray cameras. In particular, a CCD-based filmless imaging system for stereotactic biopsy procedures in mammography has been shown to significantly improve the speed and accuracy of the biopsy[8,9].

Medical X-ray imaging systems typically incorporate a cooled CCD camera system which is optically coupled to an X-ray scintillating plate. The coupling is effected by using either a lens assembly or a fiberoptic plate, Figure 2. The X-ray beam is converted into light photons by a scintillating screen; then the CCD pixel array samples the optical scene and creates a digital image. The scintillating screen is usually made of a polycrystalline material such as gadolinium oxysulfide or cesium iodide. The transfer of light from the scintillating screen to a CCD imager can be accomplished by direct contact of these components. However, the currently available CCDs are usually not larger than 5cm x 5cm. Therefore a demagnifying optical coupling (either a lens or a fiber taper) must be used between the intensifying screen and the imager in order to cover an area larger than the size of the imager. Initial demonstrations utilized a Loral 2Kx2K CCD imager with an active area of 3cm by 3cm.

Small X-ray imagers 2.6cm by 3.3cm suitable for intra-oral dental imaging are now being produced in the thousands. A scintillator plate is mounted on the face of the CCD and encapsulated in an autoclavable package for insertion in the mouth. The advantages of an all digital approach over film are similar to those in mammographic applications. A dramatic drop in radiation level is achieved. There is a significant reduction in exposure time and less X-ray neutralization of trauma drugs. The images can be displayed immediately with a wider dynamic range and can be electrically transmitted to a remote site for archival on the information superhighway.

Recently Loral has demonstrated 4Kx4K imagers with an active area of 6cm by 6cm and 8cm by 8cm. This is sufficient to image a complete biopsy area without requiring a tapered fiberoptic bundle. The scintillator plate may be attached directly to the CCD. Requirements for a full-field, 18cm by 24cm imaging system for mammographic screening and diagnosis will still require a more complex multiple CCD system.

4 HIGH RESOLUTION DIGITAL IMAGING

By the middle 80's the commercial market for camcorders was completely dominated by CCDs. Vidicon tubes common in the first generation cameras had been replaced by the solid state, stable, low power CCD. High resolution CCDs are migrating into the higher end still camera arena as a film replacement. Leaf and Mega-Vision market commercial replacement camera backs for Hasselblad cameras. The camera backs use the Loral 2Kx2K CCD for both monochrome and color operation. Their high cost (>$25K) limits their use to high end commercial photographic applications at this time. As the yields and volumes increase, prices will drop and applications expand. Table 1 outlines presently available large area imager arrays at Loral.

Loral has fabricated 4096x4096 15µm, three side buttable, scientific imaging arrays with satisfactory yields at both Newport Beach and Milpitas. A multi-phase 4096x4096 array has also been fabricated specifically for mammographic applications in both 6cm and 8 cm square formats.

The real *pièce de résistance* is a 9216x9216 pixel imaging array under development at the Loral Federal Systems 125mm wafer facility in Manassas. The pixel size is 8.75 µm square forming an image area of 8cm by 8cm. The imager uses a special three layer polysilicon process and 1.5 micron design rules to achieve optimum yield. Other large area full wafer imagers for reconnaissance applications have been demonstrated, such as a 4k by 4k[11] and a 5k by 5k[12] by Dalsa, but these are limited in size by 100mm diagonal silicon starting wafers. Our 9k imager employs the maximum area of a 125mm wafer yielding a 64 sq.cm. active area, as seen in Figure 3.

A pixel size of 8.75 microns provides a high spatial resolution of 57 cycles per millimeter. This is suitable for wide angle reconnaissance systems. Output noise is

Table 1 *Large area CCD characteristics*

ID	Pixels	Active Area	Pixel Size	Chg Cap	Outputs
CCD442A	2048x2048	3x3 cm	15 microns	100k el	2
CCD485	4096x4096	6x6 cm	15 microns	100k el	4
CCD585	4096x4096	8x8 cm	19.5 microns	170k el	4
CCD985	9216x9216	8x8 cm	8.75 microns	50k el	4

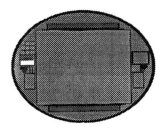

Figure 3 *Silicon wafer*

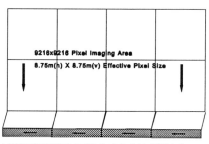

9216x9216 Pixel Imaging Area

8.75m(h) X 8.75m(v) Effective Pixel Size

4 Uni-directional Outputs per Side Readout a Total of 9216pixels

Figure 4 *Architecture of 9k CCD imager*

approximately 30 electrons with an output sensitivity greater than 3 microvolts per electron, accomplishing the ability to resolve low contrast scenes.

The current design utilizes four outputs which operate at a maximum of 15Mhz, requiring a little over 1 second to readout a single frame. A modular design allows for the simple addition of increased numbers of outputs for a higher frame rate (Figure 4). These can be added to the bottom and also the top of the array to provide a frame rate in excess of 5 frames per second.

5 CONCLUSION

As semiconductor technology continues to improve, migrating to larger and larger wafers, Charge Coupled Devices will keep pace. The use of these imaging devices will continue to move into areas previously handled only by photographic film as costs drop and speed and resolution expand. CCDs will remain a key component in imaging the visual environment.

References

1. R. A. Bredthauer, SPIE Proceedings, 1989, **1161**, 61.
2. J. C. Geary, G. A. Luppino, R. A. Bredthauer, R. J. Hlivak, and L. R. Robinson, SPIE/SPSE's Electronic Imaging Science and Technology, Charge-Coupled Devices and Solid State Optical Sensors II, 1991, **1447**.
3. C. W. Stubbs, et.al., SPIE Proceedings, 1993, **1900**, 192.

4. G. A. Luppino and K. Miller, "A Modular Dewar Design and Detector Mounting Strategy for Large-Format Astronomical CCD Mosaics", PASP, October 21, 1991.

5. G. A. Luppino, R. A. Bredthauer, and J. C. Geary, SPIE Instrumentation in Astronomy VIII, 1994, **2198**.

6. P. Suni, V. Tsai, and P. Vutz, SPIE Proceedings, 1994, **2172**, 199.

7. A. Maury, Private Communication, Observatoire de la Côte d'Azur.

8. M. F. Piccaro, E. Toker, SPIE Proceedings, 1993, **1901**, 109.

9. A. Karellas, L. J. Harris, H. Liu, M. A. Davis, and C. J. D'Orsi, *Medical Physics*, 1992, **19**, 1015.

10. H. Liu, A. Karellas, L. J. Harris, and C. J. D'Orsi, SPIE Proceedings, 1993, **1894**, 136.

11. S. R. Kamasz, M. G. Farrier, W. Washkurak, and S. G. Chamberlain, SPIE Proceedings, 1993, **1952**, 119.

12. S. R. Kamasz, M. G. Farrier, F. Ma, R. Sabila, and S. G. Chamberlain, SPIE Proceedings, 1994, **2172**, 155.

A NEW LOW-NOISE CCD FOR SPECTROSCOPY

John C. Geary

Smithsonian Astrophysical Observatory
60 Garden St.
Cambridge, MA 02138

1 INTRODUCTION

Astronomical spectroscopy, especially at high dispersion, is a demanding task for a CCD imager. Even after integration times of many minutes, flux levels per pixel may measure only a few photons. This puts a great premium on having a detector with both high quantum efficiency and very low readout noise. Until recently, there were few if any commercially available devices for this demanding application, but this may now be changing. Some recently introduced imagers have the requisite sensitivity and noise performance, with formats big enough to accommodate the focal planes of existing astronomical spectrographs and prices low enough to be affordable for even modest installations.

Scientific Imaging Technology Inc. (SITe) has recently produced several new imagers with 15-micron pixels and fairly large formats. The devices chosen for this study had formats of 1752 X 532, giving an imaging area of 26.3 X 8.0 mm. A total of four imagers were tested, one bought by SAO and three additional devices provided on loan from SITe. All are thinned for backside illumination. A liquid N2 dewar of our own construction was used to allow cooling to -100 C for testing.

A schematic diagram for this device is shown in Figure 1. Although four outputs are provided, one is an experimental and apparently nonfunctional design, and time permitted only one of the other three (the "C" amplifier) to be tested.

2 LABORATORY TESTS

Because we anticipated good noise performance from the output amplifiers provided in this design, it was essential to use a preamplifier having the requisite low noise itself. As part of an ongoing effort here at SAO to investigate preamplifier designs for low noise work and following a suggestion from P. Norgaard[1], we elected to use an array of AD745 opamps, configured with a positive voltage gain of about 20. A schematic diagram for the preamplifier array is shown in Figure 2.

The first order of business was to measure the electronic gain of each imager, allowing readout noise to be measured straight away. Because of past problems in deriving this value unambiguously using statistical methods, we insist on an absolute calibration of electronic gain using Fe55 xrays[2], a technique which also serves to demonstrate that there is good charge

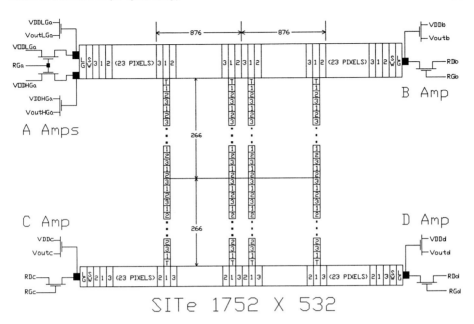

Figure 1 *Schematic Diagram for the SITe 1752 X 532 Imager. The imaging area is split in the center and is driven by two sets of parallel phases. The serials are similarly split, allowing bidirectional operation if desired. Only the C amplifier has been tested so far in our investigation.*

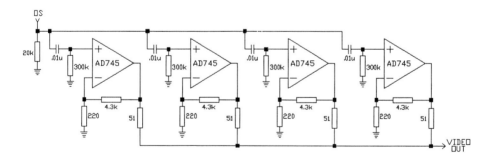

Figure 2 *Low-Noise Preamplifier Design. This circuit can be made quite compact, allowing easy mounting inside the dewar, close to the imager output.*

transfer efficiency. The latter was found to be excellent, producing sharp, unambiguous peaks for both the Fe55 alpha and beta lines at all positions on the imager.

Knowing the gain, the quantum efficiency could then be measured using an absolutely-calibrated multicolor source, an Optronics 310 system producing a known flux in discrete bands. Two of the imagers have antireflection coatings optimized for extended UV response down to the atmospheric cutoff, while the other two were optimized for the blue-visible region.

Table 1 *Operational results for the four tested 1752 X 532 imagers.*

| | Quantum Efficiency @ (nm) | | | | |
	400	450	500	600	800
1759BU03-D3 Coating: extended UV Noise: 2.9 e- Cosmetics: 2 blocked columns	54%	63%	65%	64%	53%
1759BU04-E1 Coating: extended UV Noise: 2.7 e- Cosmetics: no obvious defects	55%	61%	62%	60%	52%
1613BR06-F3 Coating: enhanced visible Noise: 4.1 e- Cosmetics: 1 blocked column		76%	75%	72%	71%	66%
1764CR01-F3 Coating: enhanced visible Noise: 3.6 e- Cosmetics: no obvious defects		70%	72%	71%	71%	63%

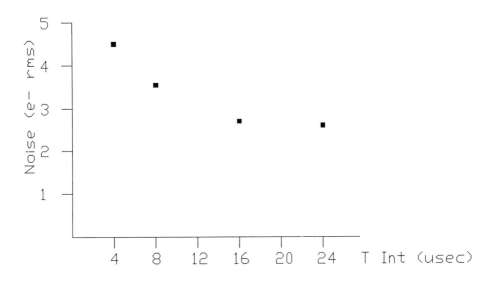

Figure 3 *Measured Noise vs. Integration Time. Time is in μsec for each half of the dual integration and does not include dead time for clocking and settling.*

The results of these measurements, along with the noise measurements, are presented below in Table 1 for the four imagers tested.

All of the noise measurements in Table 1 were taken with a dual slope integration system for signal processing. The listed rms noise is for an integration time of 16 μsec. With transfer and settling time overhead of about 6 μsec added, the resultant pixel time is then 38 μsec, or a serial pixel rate of about 26 kHz. A series of noise measurements for one of the imagers (1759BU04-E1) were made with different integration times in order to better characterize the noise spectrum of the SITe output amplifier, summarized in Figure 2. The amplifier obviously has a fairly low frequency noise corner, with white noise beginning to become dominant at higher frequencies, requiring operation at lower pixel rates in order to achieve the best noise performance.

3 TELESCOPE TESTING

The imager 1759BU04-E1 was used for one night on the SAO echelle spectrograph on our 61" telescope at Oak Ridge Observatory. We found that we could align a total of six echelle orders on the imager at once, with full coverage in the cross-dispersion direction. Due to intentional cross-dispersion astigmatism in the spectrograph (useful in olden days using photographic plates), we chose to completely collapse each order into a single serial readout in order to maximize signal. With just six serial lines of data, an added benefit was an extremely short readout time.

Weather conditions during this test were going downhill rapidly and so it is difficult at present to give an exact figure for the speed gain of the CCD over the much older photon-counting system normally used on this spectrograph. However, the observers noted that spectra were being finished so fast that they had insufficient time to log exposures by hand. In any event, it was clear that we gained at least a factor of 2-3 in speed compared to the photon-counter (peak QE = 10%). In addition, we were getting data from six orders instead of just one, which should result in better radial velocity accuracy for this particular program.

4 CONCLUSIONS

The SITe 1752 X 532 is found to be an excellent candidate for demanding spectroscopic applications, where low noise with good quantum efficiency is required. Based on a small sample, it seems that a significant number of devices perform with noise levels of 3 electrons rms or even less. This is of great importance for work at very low flux levels, and should serve as a model for future amplifier noise goals. The only drawback for the present design is the rather low-frequency noise corner, requiring readout rates that might prove troublesome, especially for larger formats and where binning cannot be used to reduce the effective pixel count.

References

1. Preben Norgaard, Copenhagen University Observatory, private communication, 1994.
2. J. Janesick, T. Eliot, S. Collins, M. Blouke, and J. Freeman, "Scientific Charge Coupled Devices", *Optical Engineering* 26, pp. 692-714, 1987.

IMAGING SPECTROSCOPY FOR INFRA-RED ASTRONOMY

C. D Mackay, I. R. Parry, F. Piché, K. Ennico

Institute of Astronomy
University of Cambridge
Madingley Road
Cambridge CB3 0HA
England

1 INTRODUCTION

Systematic surveys have now produced catalogues of over 2000 faint galaxies that cover redshifts of up to around z=1. They suggest that most massive galaxies were formed over five billion years ago and that they have changed little since then. There is, however, significant evolution in the rate of star formation in the lower mass (dwarf) galaxies which we see now as strong emission line radiation from the more distant objects. In order to get better information on galaxy evolution we need to search for and study the spectra of objects at higher redshifts where the strongest emission lines from these objects are shifted into the near infrared part of the spectrum. The objects are very faint and are just at the limit where it is difficult to separate them from the general background population of field galaxies, even with 4-metre class telescopes in the visible.

In principle, we could look to the active/adaptive optical systems that allow the atmospheric distortions to be reduced, but these faint galaxies are extended slightly even at these high redshifts to about 0.7 arcsecs at least, so such techniques will not help much here.

Working in the infrared improves sensitivity greatly because objects are easier to see against the sky in the infrared. Recent developments in near IR array detector technology give astronomers systems that are similar to the excellent ones already available in the visible. These are based on a Mercury Cadmium Telluride (MCT) layer contacted to a silicon multiplexer. They have excellent quantum efficiencies of up to 50-80 percent in the range of 1.0 to 2.5 microns, and a readout noise in a single read of under 10 electrons rms, with multiple reads further improving the read out noise. They also now have a dark current level (with the latest HAWAII 1024 x 1024 pixel devices from Rockwell) of less than 0.1 electrons per pixel per second at liquid nitrogen temperatures.

What makes things harder in the near infra-red windows (J band at 1.3 microns and H band at 1.6 microns) is the night sky emission.

2 THE NIGHT SKY

The night sky provides a uniform background illumination against which astronomers have to try to observe faint objects. Some of it comes from the upper atmosphere, so working from space will improve the near IR background. In the visible astronomers expect night sky brightness levels of 20 mag/sq arcsec in the red or 22.5 mag/sq arcsec/second in the blue, corresponding to signals of a few tens of photons per sq arcsec in relatively broad band imaging. In contrast, in the near-IR the night sky fluxes can be several tens of thousands of photons per sq arcsec in broad band (J band or H band) per second, saturating the detectors in just a few tens of seconds. This greatly limits the kinds of astronomical work that can be done.

This light arises largely from emission by atmospheric OH molecules It accounts for 98 percent of the night sky emission. in the H band and for 95 percent in the J band. Detailed spectra of the night sky shows that in both bands the emission consists of a large number (>50) of extremely narrow lines. Typical spectra are shown in Figures 1 and 2.

We could in principle observe with a very high resolution spectrograph and so minimise the influence of these lines with conventional sky subtraction techniques, but unfortunately the objects we seek are so faint that only very low resolution spectroscopy is possible if the objects are to be detected against the detector read-noise thresholds.

The instrument we are developing allows us to suppress the sky emission wavelengths without significantly reducing the flux from other wavelengths. It is based on an idea by Miahara[1] to allow the emission lines to be suppressed so as to lower the sky background by a factor of 50 (H band) and 20 (J band). Miahara has already built a system for use on the University of Hawaii 2.2 metre telescope and is now building one for the Subaru Telescope under construction on Mauna Kea, Hawaii.

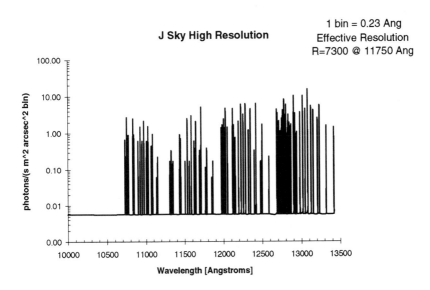

Figure 1 *The night sky spectrum in the J-band*

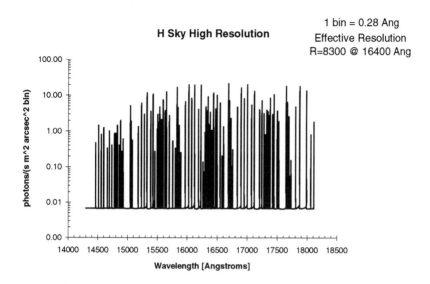

Figure 2 *The night sky spectrum in the H band*

The basic principle is to pass the light from the slit of the spectrograph through a high dispersion spectrograph that forms a high resolution spectrum on a mask that cuts out light only at the wavelengths of the emission lines. The rest of the light is reflected back through the high resolution spectrograph so that the light is recombined to produce a white light image that is now free from the emission line signals. The reconstructed white light image of the slit may then feed a much lower resolution spectrograph that is appropriate to the objects being observed.

The gain in limiting magnitude of such an OH suppression instrument is dependent on the suppression factor G and the suppression instrument efficiency E. The gain in signal-to-noise ratio (SNR) is ÖGxE. For E=0.33 this is x 3.6 in H band (or 1.4 mag) and x 2.6 in J band (or 1.0 mag). These factors are extremely significant to astronomers. A factor of 3.6 gain in SNR is achieved by replacing your old 4-metre telescope with one of 15 metres diameter. It will also allow faint objects to be seen at about three times the current maximum distance, clearly of great cosmological significance.

3 CAMBRIDGE OH SUPPRESSION INSTRUMENT (COHSI)

The instrument we are building is fed by a fibre bundle that transfers the conventional spectrograph slit at the focus of the telescope to the OH suppression instrument located on the floor of the telescope. The fibres are each fed by individual lenslets that allow 100 percent of the light falling onto the focal plane to enter the fibres, avoiding the losses usually associated with the cladding and packing of multiple fibres. The output of the fibres is arranged in a line at the focus of a large primary mirror that collimates the beam

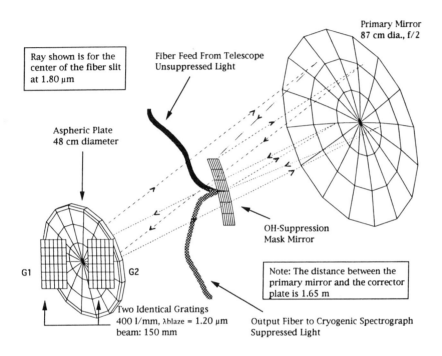

Figure 3 *The general layout of the Cambridge OH Suppression spectrograph (COHSI)*

onto the first grating (Figure 3). The grating disperses the light and reflects it back onto the primary mirror which forms a spectrum onto the OH-suppression mask mirror. The mirror only reflects those parts of the spectrum that are free from OH emission features back along a similar path, to the primary mirror and onto a second, identical grating symmetrically on the other side of the optical axis from the first. The second grating in turn undisperses the light (now with the OH emission lines removed) and forms an image of the input fibre slit next to it. This OH suppressed slit may then be used to feed a low resolution instrument.

We are also building a cryogenic spectrograph to work in conjunction with COHSI. It will allow us to observe simultaneously in both H (1.0-1.35 microns) and J (1.44-1.8 microns) band using two NICMOS III detectors, each of 256 x 256 pixels of 40 microns square.

The combined instrument design consisting of the OH suppression unit and the cryogenic spectrograph predicts a relative E value of 0.35, and an overall throughput from the top of the aperture to detected electrons of about 15 percent.

4 FEEDING COHSI FOR SPECTROSCOPY AND IMAGING

We will use fibres fed with miniature lenses in an array as the input to COHSI. In fact as the fibres are independent there is no reason to restrict operation to slit inputs. If, for example, we wish to look at the spectrum across a two-dimensional object then we may arrange for the lenslet-fed fibres to cover the object in question. We can also arrange for several fibres to be placed well away from the object for reference since we will still wish to use night-sky subtraction techniques to remove the residual sky light.

In addition we do not need to pass the light that comes out of COHSI with the OH lines suppressed into a second spectrograph. Should we wish to simply do very faint sky limited imaging then the COHSI output fibres may be imaged directly through passband filters.

The COHSI design will allow simultaneous suppression of the OH emission in a maximum of 1600 fibres. The size of the individual lenslets allows a trade-off between the resolution of the work and the overall field size covered.

ACKNOWLEDGEMENTS

This work is funded partly by a grant from PPARC no. GR/J79300. We wish also to thank Dr A K Velan for generous support of this project.

References

1. T. Miahara et al., Proc SPIE, 1993, 1946, 581.

Dr C D Mackay:
tel: +44-1223-337543, fax: +44-1223-337523, e-mail: cdm@ast.cam.ac.uk

LOW-NOISE CCD IMAGING AT KILOHERTZ FRAME RATES

C. D. Mackay

AstroCam Ltd
Innovation Centre
Cambridge Science Park
Milton Road
Cambridge CB4 4GS
England

1 INTRODUCTION

There are a great many applications that require high speed imaging. Traditionally, users have accepted that the image quality obtained from systems intended for high speed use was much poorer than that obtained with precision slow scan scientific CCD cameras in terms of readout noise, low contrast capability and low light level sensitivity. Recently, however, CCD controllers have become available from companies such as AstroCam Ltd that are able to offer very high levels of performance with specific CCDs. These have the potential to offer many researchers much better imaging system performance than has hitherto seemed possible.

Two examples of programmes that have already benefited from these developments will be described, followed by a discussion of the design aspects that are critically important to such systems.

2 ACTIVE AND ADAPTIVE OPTICS SYSTEMS FOR ASTRONOMY

Ground based astronomy is seriously hampered by turbulence in the atmosphere. The wavefronts coming from a star are distorted so that the images formed are smeared out to a resolution of typically one arc second in diameter. The telescope is capable of forming images perhaps 40 times smaller if no such phase errors were introduced by the atmosphere (for a 4 metre diameter telescope). Systems have been designed and built to overcome these limitations in the following way. The main requirement is to sense the phase errors by looking at the image distortions quickly so that a correcting element may be made to distort to compensate for these errors. The system has to work quickly (10 msec is the typical atmospheric correlation time in the visible), and it must work with the lowest possible signal levels.

Under typical conditions in the visible, the atmosphere shows relatively constant phase gradients across patches of 10-20 cm diameter. There will be typically a linear phase gradient across patches of up to 50 cm. On larger scales, the phase errors become more complex. If the pupil of the telescope is re-imaged so that it falls on a lenslet array, then each lenslet forms its own image of the star with the light from the patch of the pupil it covers. By studying the relative motion of the multiple sub-images, we can establish how the atmospheric phase distortions are changing with time. This arrangement is often referred to as a Shack-Hartmann wavefront sensor.

The atmosphere changes quickly, and the turbulence has different characteristics at different telescopes on different sites. If an effective AO system is to be built, the designer needs to know a lot about local conditions to be contended with. The UK AO programme has funded the development of a high speed imaging system that uses a lenslet array of 8 x 8 elements imaged onto a high speed Capella camera system built by AstroCam Ltd, Cambridge, England. It uses a custom 64 x 64 frame transfer CCD with 15 x 15 micron pixels designed by John Geary (Smithsonian, USA) and manufactured by Loral (USA). The CCD is unthinned (front illuminated). The Capella system reads out images at a maximum rate of 865 frames per second at full resolution into a SparcStation computer. Data are taken at up to 5.5 MHz pixel rate with 12-bit accuracy (the Capella is capable of 14-bit accuracy at up to 5 MHz) at up to 32000 frames per run (set only by the memory capacity of the computer) before saving to disk. The data allow a detailed analysis of the motions of the star images from each of the 50 cm x 50 cm patches of the WHT 4.2 metre telescope on La Palma. The high maximum frame rate allows the movements from the atmosphere to be frozen and followed reliably. For even higher time resolution, the Capella may be programmed to read out a sub-array and give yet higher frame rates (such as 1200 Hz at 32 x 64 pixels).

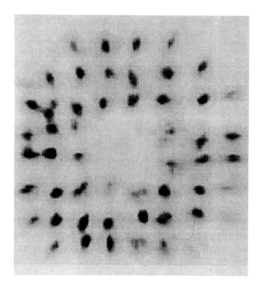

Figure 1   *A typical image from the Shack-Hartmann wavefront sensor installed on the 4.2 metre William Herschel Telescope on La Palma. The distortions from a rectangular grid are partly atmospheric, partly due to telescope effects (particularly the mess at the 10 o'clock position), and partly due to optical geometric effects in the experimental set-up. The data were taken of the star Alpha Bootes in the red at 865 frames per second.*

The system is now in routine use as a seeing and site evaluation monitor on the 4.2 metre telescope. The scales (both temporal and spatial) of the turbulence increase as the working wavelength is increased, both by the power 1.2 of the wavelength. This makes it possible to work much more easily if the goal is for measurement and phase error correction to be effective only in the near infrared rather than the visible. The performance of the wavefront sensor is the most critical component in determining the limiting sensitivity of the system, since if the target star is not bright enough to give a good signal in the correlation time over the mono-phase aperture then the system will not work. Telescopes are usually limited in sensitivity by other things and allow integrations of minutes or hours using the whole telescope rather than just a tiny sub-aperture.

3   STUDIES OF MAMMALIAN NEURONES

There is increasing interest in the study of mammalian neurones. The neurones work with complex electrical activity patterns that may be visualised with voltage sensitive dyes or fluorophores. These dyes are affected only rather slightly by the voltage changes so any attempt to measure electrical potential patterns in neurones needs systems with excellent low-contrast capabilities as well as good read-out speed.

AstroCam Ltd has developed a fast read-out system that uses a 512 x 512 frame transfer CCD manufactured by EEV (UK) with 15 micron pixels. The system uses a Capella camera system manufactured by AstroCam Ltd, Cambridge, England that reads out full resolution images at 17 Hz, digitised to 14-bit (16000 grey level) accuracy, and much higher frame rates if the image is binned and/or a sub-array is selected for read-out.

The normal mode of operation has the user taking full frame images so as to identify the principle area of interest and then reading that area as quickly as possible as the neurones are stimulated electrically. The data must be captured and stored at high speed, a process that requires specialist hardware and software solutions if the system is to be efficient and effective. Software must also allow easy analysis of what are often very large three-dimensional data sets if the results are to be made available to the research worker immediately after a data recording run is complete. These facilities are provided by the support hardware and software that is part of a full AstroCam Capella Imaging System.

4   HIGH SPEED CCD ARCHITECTURES

It is inevitable that the readout noise is poorer when operating CCDs at high speed than when operating them at slower speeds. At a given frame rate, the actual pixel rate may be reduced by using CCD architectures with multiple outputs.

The simplest way of achieving this is to split the output register, place an output amplifier at each end, and allow the option of clocking the CCD to one or the other or both amplifiers at once. This is the architecture used by EEV(UK) and SITe (USA) for their new families of three-edge buttable large area CCDs. The parallel imaging area can also be split, and separate output registers provided at both top and bottom of the imaging array again split to give four parallel outputs. This is the architecture used by SITe with their 1024 x 1024 and 2048 x 2048 CCDs.

An alternate way of creating four outputs is to split the output register into four (or more) sections, as is done by Thomson (France) with their 3-edge buttable 2048 x 2048 pixel THX7897 device. Yet another way is to have multiple output registers that lie alongside one another and may be clocked in parallel, as is done by Kodak with their 2048 x 2048 pixel KAF-4200 device. A more extreme solution is to use one output amplifier per column as in the 64 x 64 CCD originally built by Tektronix (now SITe) to give frame rates in the tens of kilohertz range.

5   HIGH SPEED CCD READOUT NOISE

The traditional view is that low read out noise (<10 electrons rms.) is only possible with slow readout (<100 KHz), and that MHz pixel rates inevitably give readout rates in the 50 plus range. There are still many CCDs for which this is true; these are sold for slower read rates. However, there are an increasing number of devices that perform rather well at higher readout speeds.

One critical component in the CCD design is the on-chip amplifier. Traditional single transistor designs need to be buffered so that higher gains may be realised and so that finite capacitances may be driven without the gain being pulled down.

Considerable progress has been made in building high-gain amplifiers that give good readout noise even at high speed. Kodak have been leaders in this area for a number of years, and essentially all their CCDs have high-speed and low-noise simultaneously. Only recently have other companies been able to improve on the Kodak figures. Representative results are shown in Figure 2. In this figure the data on the Kodak KAF-1000 are representative of what has been available with Kodak CCDs for several years. Readout noise levels of under 10 electrons rms. have been routinely achieved by AstroCam's Capella camera systems for some years at pixel rates of 1 MHz, with levels of 25-50 electrons at 5 MHz. The data in Figure 2 also show the improvements much more recently achieved by EEV with their latest architectures of readout amplifiers. The latest results on the new EEV CCD42 (described by Jorden and Oates elsewhere in these conference proceedings) show the sub-2 electron read noise predicted in slow-speed, making it highly likely that the predicted 4-5 electrons read noise at 1 MHz will be achieved.

6   OTHER CONSIDERATIONS FOR HIGH SPEED CCD DESIGN

There are a number of other factors that must be considered when high speed operation is to be combined with low readout noise and the widest dynamic range.

As CCD areas get larger, the parallel transfer times get longer because of the increased capacitances of the image transfer electrodes. This can force designers to smaller CCD pixels (and this is the way Kodak have gone with 9 micron pixels for their largest arrays). This can reduce the CCD full well capacity of the device, but already the performance of some large area CCDs makes it clear that a dynamic range of 14 bits is achieved at 1 MHz and 13 bits at 5 MHz. Kodak 9 micron pixel CCDs give an 85 000 electron full well, and the Kodak KAF-1000 with 24 micron pixels has a full well in excess of 200 000 electrons, or 20 000 times the readout noise at 1-2 MHz pixel rate.

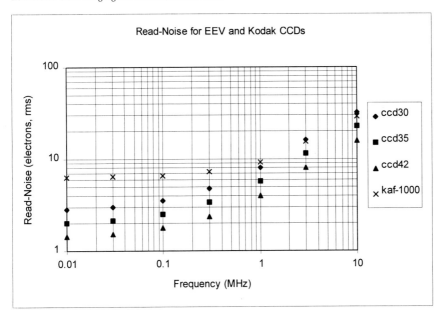

Figure 2 *The readout noise levels achieved for a selection of CCDs. The ccd30, ccd35 and ccd42 are different generations of EEV CCDs, while the kaf1000 is a large pixel CCD from Kodak (USA).*

One important aspect of the design of multiple output CCDs for slow or fast readout applications is the problem of cross talk between the several channels. For high quality scientific applications, a significant detection of a real object covering 25 pixels with a CCD that has a 10 electron read out noise might be as low as 150-200 electrons total or only 6-8 electrons per pixel. A CCD design that has common supply lines with finite impedances at high speeds to the transistor output drains, for example, could easily impose a crosstalk from a channel that is near saturation that will be easily detectable in channels with weak signal in them.

High speed camera systems such as the AstroCam Capella offer performance now that for most applications is indistinguishable from the performance usually associated with much slower camera systems. With the next generation of multiple output CCDs, even faster performance will become routinely available.

Dr C D Mackay:
tel: +44-1223-420705, fax: +44-1223-423021, e-mail: cdm@astrom.co.uk

THE CCD DETECTORS OF THE HALLEY MULTICOLOUR CAMERA AFTER SEVEN YEARS IN FLIGHT

J.R. Kramm, H.U. Keller and N. Thomas

Max-Planck Institut für Aeronomie,
D-37189 Katlenburg-Lindau, F.R.G.

1 INTRODUCTION

The Halley Multicolour Camera (HMC) on board ESA's spacecraft Giotto used TI uniphase CCD detectors in its focal plane. The spacecraft was launched in July 1985 and successfully completed a fly-by of comet Halley in March 1986. Subsequently, the spacecraft was re-targeted for an encounter with comet Grigg-Skjellerup in July 1992. Although the optics of the camera were damaged during the Halley encounter, the CCDs and electronics survived and remained fully operational. Tests on the CCDs were performed in May 1990 and July 1992, five and seven years, respectively, after launch. Changes in the performance of the detectors occurred during cruise and are reported here.

2 HMC CCD DETECTORS AND DETECTOR OPERATION SCHEME

The Halley Multicolour Camera (HMC) on-board ESA's spacecraft Giotto passed comet Halley in March 1986 after being in space for nearly 9 months. The camera, a Ritchey-Chrétien type telescope, was designed to track and image the comet on a spin-stabilized spacecraft moving at about 68 km/s relative to the comet. HMC provided impressive pictures, which clearly showed the shape of the cometary nucleus and a number of structures on the surface. A full instrument description has been given by Schmidt et al. (1986) and Keller et al. (1987), and images were published by Keller et al. (1988), Thomas et al. (1988) and Keller et al. (1995).

A time-delay and integration (TDI) technique was introduced to yield high resolution images from a moving target. Two CCDs were used to obtain four identical sections all covered with metal masks so that no light could penetrate. As shown in Figure 1, a small slit above the top lines of each of the four sections was left open for exposure, while all of the rest was used only for intermediate storage. When the image passed across the slit, the CCD had to be clocked with a synchronized clock frequency which matched the geometrical charge transfer with the image speed across the detector. The generated charge was accumulated below the mask and temporarily stored without clocking until all commanded sensors were exposed and the mechanical shutter could be closed. The two CCDs were read out in parallel during the remainder of the 4 seconds spin period. Three of the four slits were covered with different optical filters while the main detector slit was equipped with a filter wheel containing 11 different filters including a clear filter.

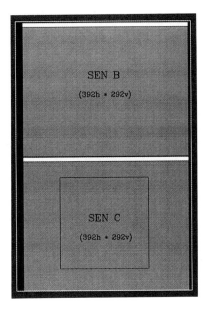

Figure 1 *Schematic diagram of an HMC CCD detector with metal mask, 2 imaging slits, intermediate storage registers and covered reference pixels on both sides. One CCD detector was split into two sub-sensors. The masked window in the sensor C area indicates a 196h x 196v sub-area cut-out used in some of the later investigations.*

In the reference position, this filter wheel was opaque and was used as the optical shutter during the readout of the CCD. A certain delay (a few 100 ms) was needed to close the shutter from any filter position. The delay time was associated with a significant build-up of dark charge. (Note that this delay corresponded to a period of time without parallel clocking and was therefore equivalent to the exposure time in standard CCD camera applications.)

The camera electronics were developed and built in the time between 1982 and 1985. Due to the tight delivery schedule, the risks had to be minimized. Commercial area CCDs manufactured by Texas Instruments for television cameras were selected. For the HMC application, some minor modifications were introduced to improve the quantum efficiency mainly by removing the antiblooming structure.

The CCDs consisted of two sections of 292 lines of 390 pixels each. The pixel size was 22.4 μ square. The registers were made for a uniphase clocking technique with inverted mode to yield lower dark current. The clock amplitudes had to be about 17 volts, well adjusted to obtain a compromise between dark current reduction and spurious charge generation. Spurious charge was always a problem with these uniphase devices. A versatile mixed mode clocking had to be introduced to obtain an optimal balance between two level clocking (high spurious charge, up to 10% full well) and three level clocking (high dark current). All efforts to keep the dark current low became very important, because the operating temperature on the spacecraft was always higher than predicted.

The typical output amplifier responsivity was 0.5 μV/e⁻. The output amplifier consisted of a two stage Darlington FET configuration with typical noise figures of 100 to 200 e⁻ (TV application). With some external modifications, a readout noise of 25 to 35 e⁻ could be obtained for all HMC flight candidates (Kramm et al., 1985).

Each pixel was read out in 13.75 μs and converted with a 12 bit analog to digital converter. An additional gain switch was included to enlarge the dynamic range by a factor of four. Gain 1 was mainly used when reading binned superpixels. Single pixels were mostly read in gain 4 position yielding a sensitivity of 37 e⁻/DU.

A square root data reduction scheme was implemented because of the limited telemetry rate. Due to this non-recursive compression technique, some resolution loss is visible during the mapping of hot pixels with high amplitudes. More data reduction could be obtained by reading only defined sub-area regions (windows) or by binning several pixels into one superpixel.

On the spacecraft, the CCDs were mounted with the columns perpendicular to the spin axis of the spacecraft. The focal plane of HMC was split using a prism-shaped mirror. A CCD was mounted on each side of the split focal plane. During the rotation of the camera, the serial registers moved on a plane parallel to the spin axis. The CCDs were mounted on aluminium blocks, 10 mm thick, in a cylinder housing made of 6 mm titanium. The cylinder was covered on both ends with 3 to 4 mm aluminium. The front end and the back of the spacecraft consisted of 40 mm Al honeycomb (Hartwig).

A passive thermal cooling of the detectors was adopted in accordance with the spacecraft power restrictions. During operation, a rise in the temperature of the CCDs was expected due to the electrical power dissipation of the detectors and the electronics near the detectors.

The CCDs were manufactured in 1983, implemented and tested in 1984 and, finally, launched in 1985. During the cruise, a large number of tests were performed to obtain data for dark current and responsivity matrices and optical performance evaluations. Dark matrices were collected over a wide range of operating temperatures.

The dark matrices were influenced by several parameters, such as temperature, TDI clocking speed, superpixel format, individual filter wheel delay and the gain switch position. The wide variety of impacts of all these parameters meant that individual dark matrices for each of the possible combinations of parameters could neither be taken nor stored. Calibration matrices had to be constructed. A detailed understanding of the response of each individual pixel was mandatory in order to calibrate the resulting data from the comet fly-by.

During the encounter, more than 2200 images were obtained, most of them (nearly 99%) with the main detector, SEN C, the lower section of CCD 1. Consequently, the individuality of the pixels on this detector was very well mapped. The following analysis, therefore, will focus on this detector only.

Table 1 *Performance changes on Detector C*

Date	Temperature	Data
July 02, 1985		Giotto launched
1. March 04, 1986	-4.5 ... -1.7 C	196*h* x 196*v* sub-area
2. March 13, 1986	-7.5 ... -4.5 C	Encounter, all formats
3. May 05, 1990	(+28 ... +29 C)	390*h* x 292*v* full frame
4. July 12, 1992	-2.4 ... -2.1 C	390*h* x 292*v* full frame

3 AVAILABLE DATA FOR PERFORMANCE COMPARISONS

In the months before the encounter, test data with detector temperatures between -20 C and +10 C were received. During encounter, the temperature increased continuously from -7.5 C to -4.5 C within three hours.

Despite the large temperature range, most of the pixel inhomogeneity apparently depended purely on the detector temperature. However, several hot pixels disappeared with additional cooling and instantly came back after warming up again. Our hot pixel maps had to be spaced in 1 C increments. Typically, the number of hot pixels with extra charge of >1000 e⁻ increased by 20 to 30% per degree. Consequently, confident performance comparisons could be made only on datasets received from operations taken under similar temperature conditions.

Data from the following passes recorded under comparable temperature conditions were available to identify performance changes on detector C in the space environment (Table 1). The data received in May 1990 and June 1992 were unsuitable for comparisons of the responsivity matrix. Some images contained exposures with stray light from the Sun but they were taken with lower resolution by using superpixel format $4h \times 3v$ pixels per pixel.

3.1 Orbit of Giotto during cruise

The orbit of the Giotto spacecraft after launch took the spacecraft initially inside the Earth's orbit to 0.72 AU from the Sun on December 31, 1985 before drifting outwards again for its encounter with comet Halley 0.89 AU from the Sun on March 14, 1986. Shortly after the Halley fly-by, Giotto's trajectory was modified so that an Earth swing-by occurred on July 2, 1990. After this Earth gravity assist, the spacecraft entered a low eccentricity, heliocentric orbit which was mostly outside the orbit of the Earth (semi-major axis = 1.08 AU). The encounter with comet Grigg-Skjellerup occurred on July 10, 1992 at a heliocentric distance of 1.01 AU.

4 PRE-ENCOUNTER DETECTOR PERFORMANCE STATUS

Pixel inhomogeneity was permanently observed. In the temperature range revealed during encounter, some pixels had an extra charge of up to 1200 DU ($\approx$ 45 000 e⁻). Different effects were separated.

The effects were classified into the following groups:
    a. Random extra charge due to cosmic ray events
    b. Stable hot pixels with constant charge generation
    c. Hot pixels with delayed or limited charge generation
    d. Bad pixels with unstable extra charge generation

4.1 Random extra charge due to cosmic ray events

Significant extra charge distributed randomly on sensor locations was interpreted as being due to cosmic ray events. One or more pixels could be affected. In some images up to four different events were detected. In one image we found 48 adjacent pixels with extra charge from one event.

Figure 2 shows a plot of extra charge (charge above regular dark current) in one particular pixel extracted from the 131 encounter sequence of the full frame images. This pixel received extra charge from a cosmic ray event only once during encounter. After the charge had been read out of the CCD, no significant residual was left in the pixel as further readouts demonstrated.

The average lifetime of each pixel--the time between the start of the exposure and reading the centre pixel--on detector C was $t_i \approx 1.2$ s. Every second image contained at least one event. Averaged over the 131 full frame encounter images, 3.38 pixels (of 113 880 pixels total) in each image were affected by cosmic rays.

Consequently, at a heliocentric distance of 0.89 AU, the observed rate of the number of cosmic events was 1.75 cm^{-2}s^{-1}.

4.2 Stable hot pixels with constant charge generation

Individual pixels had a significantly higher dark current generation which introduced a constant dark charge over time. Due to the TDI operation scheme, the extra charge of such a defective pixel affected the whole column.

- Each pixel **below** the defective pixel got extra charge during the TDI illumination depending on the actual TDI clocking timing (14 μs up to 1000 μs/line).
- The extra charge **within** the defective pixel depended on the delay time without clocking needed to close the shutter. For the clear filter, 270 ms were needed to turn the filter wheel back to the shutter position.

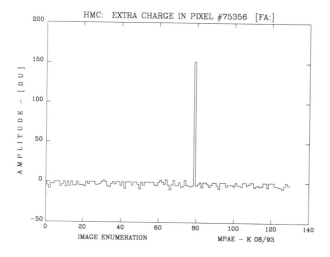

Figure 2   *Extra charge in a single pixel location from a cosmic ray event; except for one image, all other images returned only minor noise contributions*

- Each pixel **above** the defective pixel got extra charge during the readout of the data. The time needed to read one line was for 5.4 ms.

Figure 3 shows an image and a plot of the indicated column which contains a stable hot pixel (see indicating arrow). As has been mentioned before, the additional charge in all pixels below and all pixels above the hot pixel depended strongly on the individual timing used to make the image. With care, entire columns with extra charge from hot pixels with constant charge generation can be restored.

4.3 Hot pixels with delayed charge generation

Some individual pixels seemed to start as regular pixels but, with some delay, they began to accumulate higher dark current. They introduced extra charge mainly into one individual pixel location. These pixels did not add significant charge to the rest of the column. Hot pixels with delayed charge generation could be calibrated.

Other pixels had a limited extra charge generation which ended when a certain charge was introduced into the pixel location. They appeared more or less like pixels with a constant additional charge. An example of such a pixel is plotted in Figure 4.

The extra charge in this pixel was extracted from the encounter data with clear, red, orange and blue filter exposures with shutter delay times of 270 ms, 270ms, 220ms and 320ms, respectively. The data amplitudes were compensated for the delay time differences. A strong regular pattern correlated with the filter sequence dominated (see Figure 4a). Without any delay time compensation, a much smaller pattern was obtained (Figure 4b). Consequently, in the described timing range, the amplitude did not depend on the delay time. This pixel always had about 1 90 DU extra charge over a wide temperature range--and for

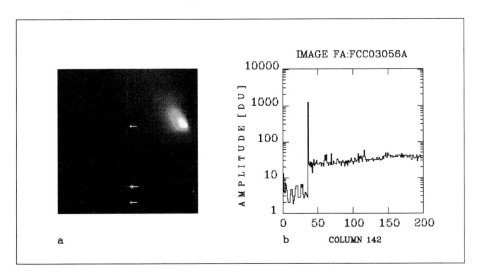

a   b   COLUMN 142

Figure 3   *Column with a stable hot pixel appears like a slightly brighter column in an image with one bright spot (a); plot of extra charge introduced into the column (b)*

years. The remaining spikes in Figure 4b indicated some hypersensitivity, because for clear filter images, the amplitude was slightly larger than on average. (This section of the CCD was exposed by flat coma light during encounter; a similar but more flat amplitude could be obtained from dark images.)

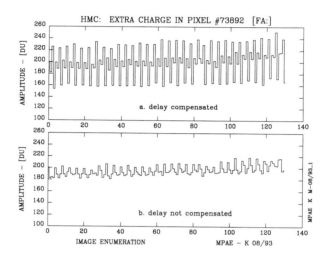

Figure 4   *Extra charge in a pixel with constant add-on charge with filter delay time compensated (a) and without delay time compensation (b)*

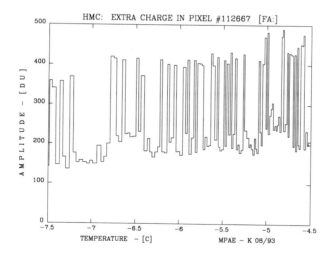

Figure 5   *Bad (metastable) pixel #112667 with short time intensity variations plotted over temperature. During exposure of these 131 images, the detector temperature increased continously from -7.5 C to -4.5 C.*

4.4 Bad pixels with metastable extra charge generation

Pixels with a metastable extra dark current generation were also found on the detectors. They flipped between two or more different levels of additional charge generation seemingly without any reason and independent of certain temperature changes.

Figures 5 to 7 show three examples of metastable charge generation taken again from the 131 full frame encounter images. Different filter delay times were normalized to the delay for clear filter exposures to obtain comparable conditions. Here, the extra charge was plotted against the temperature. Apart from a small increase in the average intensity due to the rising temperature, no significant dependance of the flipping activity could be found on metastable hot pixels in this temperature range. Attempts to calibrate such metastable pixels in an unknown image could be very critical.

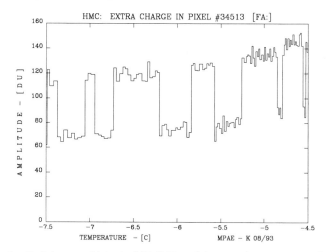

Figure 6 *Bad (metastable) pixel #34513 with longer changing periods*

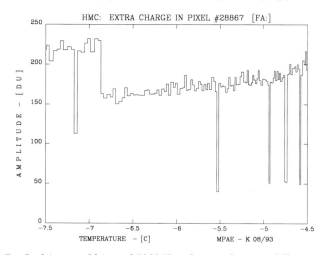

Figure 7 *Bad (metastable) pixel #28867 with more than two different extra charge generation rates*

5 SUMMARY OF THE PRE-ENCOUNTER DETECTOR STATUS

The dark current was extremely low compared to other detectors available at the time of sensor selection. On average, only 400 e⁻ to 700 e⁻ were obtained in each pixel during encounter. At a temperature of -7.5 C, about 190 hot pixels with amplitudes of >30 DU (≈ 1125 e⁻) were registered. At the end of the encounter, the number had approximately doubled when the detector temperature had increased to -4.5 C.

On the main detector, three bad (metastable) pixels were located which generally had to be removed by interpolation. Fortunately, they were situated in regions of little scientific interest.

At the end of the approach to comet Halley, Giotto was hit by cometary dust particles (see Curdt et al., 1988). The camera electronics were reset and imaging ended. Further investigations showed that the optical system was severely damaged. The CCDs and the electronics, however, could be restarted and operated nominally.

6 POST-ENCOUNTER DETECTOR PERFORMANCE STATUS

The Giotto spacecraft was re-activated successfully in May 1990 and July 1992. Some restrictions due to further limitations in the power supplied by the spacecraft solar cells were necessary but HMC could be fully tested.

Extensive investigations on the light sensitivity of the detectors could be performed successfully in 1990 on images exposed with stray light from the Sun. The CCDs were still sensitive to light but the optical path was blocked permanently. Hence, HMC could still be used to investigate the state of dark images from the detectors.

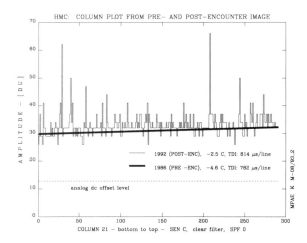

Figure 8 *Plot of column 21 of detector C, 1992 data, and same column averaged over the encounter data, 1986*

The analogue dc offset level monitored during the readout of the CCD at the beginning of each line was still nominal. This allowed direct comparisons between the data.

The regular dark current on undistorted pixels was compared. Figure 8 shows a plot of column 21 of 1992 data. For comparison, the average slope of the regular dark current of the same column taken from the encounter data has been overplotted. The timing parameters on both images were similar, only the detector temperatures differed by 2.1 C. The observed difference of 2 to 3 DU agreed very well with the results obtained during calibration of this detector in this temperature range. We conclude that after seven years in space there was no significant change in the mean dark current level.

The structure of hot pixels appeared to be completely different. The "old" well-known hot pixels mostly disappeared or continued to respond with much lower amplitude. On the other hand, more new hot pixels came up, quite a few of them with metastable extra charge generation. The former dark current matrices were obsolete.

In Figure 9, cumulative histograms of the hot pixel extra charge amplitudes from earlier and from later data have been plotted. The data were obtained from co-aligned 196*h* x 196*v* sub-area readouts covering 34% of the full frame image area. The intensity of hot spots had obviously changed.

A typical dark image from the 1992 data with hot spots and vertical structure is shown in Figure 10. Data manipulations were needed to reduce the wide dynamic range. The data were clamped to 1200 DU and printed with an exponent of 0.25.

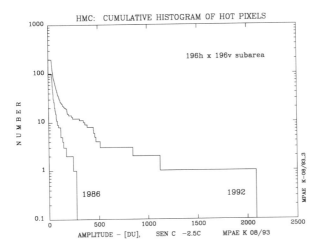

Figure 9 *Cumulative histogram plots of hot pixel extra amplitudes on detector C, 196h x 96v, 1986 encounter data and 1992 data*

Figure 10 *Typical dark current from a full frame image of 1992 data*

A typical change in the pixel performance could be observed in column 83 containing in 1986 one stable hot pixel of about 930 DU extra charge, see Figures 11 and 12. Six years later, the extra charge generation was diminished to 200 DU, but in the same column two new hot pixels appeared, one with a stable charge generation rate of up to 1240 DU. Similar effects could be found in other columns.

The three metastable pixels which were always present in the 1986 data had, six years later, some extra charge but much lower generation rates. In the 1992 data, they appeared like moderate hot pixels. A comparison between the extra charge generation rates in 1986 and 1992 is given in Figure 13.

On the other hand, pixels which became metastable pixels after six years in space, started as regular pixels without any peculiarities. A plot of the typical extra charge generation in such a pixel is printed in Figure 14.

7 SUMMARY

The Halley Multicolour Camera (HMC) on-board the European Space Agency's Giotto spacecraft used two uniphase CCD detectors produced by Texas Instruments. The spacecraft was launched in 1985. The camera was operated from October 1985 for testing and again during an encounter with comet Halley in March 1986. The camera was re-activated in May 1990 and July 1992.

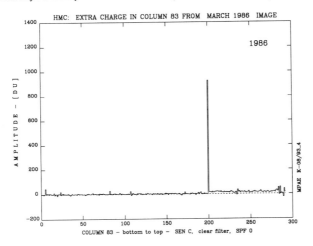

Figure 11 *Plot of extra charge in column 83, 1986 encounter data with one stable hot spot*

The main sub-sensor used was detector C and was consequently very well mapped. Comparable data were obtained for detector temperatures of -4.6 to -2.5 C. Performance changes during the cruise of 6 or 7 years have been investigated.

The results can be summarized as follows:

- Dark current on regular pixels remained constant.
- Cosmic ray events at a rate of 1.75 cm^{-2} s^{-1} were obtained at the heliocentric distance of 0.89 AU in March 1986. Residual effects could not be observed.
- The number and the intensity of hot pixels increased significantly. On a 196 x196 sub-area, the number of hot pixels with >820 e^- extra charge increased from 103 (1986) to 198 (1992). Up to 120,000 e^- of extra charge were found in a single pixel. On average each column had at least 1 hot pixel. Nearly 50% of the

  hot pixels were stable and also introduced extra charge into the corresponding column. The remaining hot pixels were of delayed type of extra charge generation and produced additional charge only in one pixel location.

- The vertical structure introduced by stable hot pixels changed in accordance with the hot pixel intensity. A strange column pattern up to 4500 e^- per pixel was observed depending on the timing parameters of the camera. Nearly half of the columns were affected.

- The performance of defective (metastable) pixels changed completely.
    - The three metastable pixels in 1986 appeared as moderate hot pixels in 1992.
    - About 10 metastable pixels were located in the later data, which started with no significant individuality in 1986.

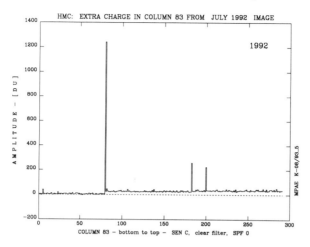

Figure 12 *Two new hot spots appeared in 1992 in column 83, one with 1240 DU extra charge generation.*

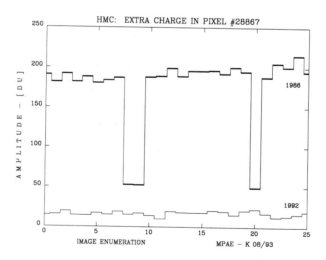

Figure 13 *A metastable pixel has been changed to a moderate hot spot after six years in flight.*

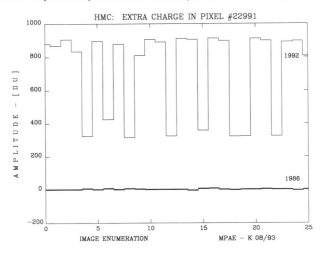

Figure 14 *A pixel with metastable charge generation in 1992 started with no significant individuality in 1986*

- Pixels with a constant add-on extra charge were found. One pixel, as an example, in the temperature range between -7.5 C and +2.5 C generated permanently an extra charge of 7100 e⁻. The charge did not depend on the shutter delay time (which corresponded to the exposure time in standard CCD camera applications).

Calibration matrices and bad pixel maps have to be rebuilt to calibrate the detector performance at each phase of the mission.

References

1. W. K. H. Schmidt, H.U. Keller, K. Wilhelm, et al., eds. Reinhard, R.& Battrick, B., *ESA SP-1077*, 1986, 149.
2. H. U. Keller, F. Angrille, C. Becker, et al., *J. Phys. E*, 1987, **20**, 807.
3. H. U. Keller, R. Kramm and N. Thomas, *Nature*, 1988, **331**, 227.
4. N. Thomas, J. R. Kramm, and H. U. Keller, *Dust in the Universe*, 1988, eds. Bailey, M.E. & Williams, D. A., 161.
5. H. U. Keller, W. Curdt, J. R. Kramm, and N. Thomas, *ESA-SP 1127*, 1995, eds. Reinhard, R., Longdon, N., 1.
6. J. R. Kramm, and H. U. Keller, *Adv. Electronics Elec. Phys.*, 1985, **64A**, 193.
7. H. Hartwig, private communication, Max-Planck-Institut für Aeronomie, D-37189 Katlenburg-Lindau, F.R.G.
8. W. Curdt and H. U. Keller, *ESA Journal*, 1988, **12**, 189.

ADVANCED IMAGING SYSTEMS FROM XEDAR CORPORATION

Gregg Herbison

Xedar Corporation
2500 Central Avenue
Boulder, Colorado 80301

1 INTRODUCTION

Photon imaging has played a major role in science and discovery for many decades. Since the invention of the Charge Coupled Device (CCD) in the early 1970's, the flexibility and number of applications for instruments using this technology has flourished. However, extracting the absolute maximum performance from these devices has become a challenge as the diversity of applications and performance requirements is expanding.

Since its inception in 1974, Xedar Corporation has been assisting the scientific and industrial community to create application-specific instrumentation using the most recent advancements in CCDs. Together with new developments in CCDs, Xedar is pursuing new methods in configurations, both electronically and mechanically, to keep pace with the expansion of applications requiring the use of CCDs.

2 INNOVATIVE MECHANICAL CONFIGURATIONS

The key to any imaging application is getting photons to the CCD. Not only are the number of photons critical, the wavelength is also a prime consideration. Most silicon based CCDs have a spectral responsivity range from 400 nm to 1100 nm with their peak detectivity near 700 nm. For many applications within this range, intensification may be required as the number of available photons are not sufficient to provide the required signal to noise ratio. In other applications, the spectral energy available may not be within the responsivity range for a silicon based CCD and may require a wavelength conversion medium, such as a phosphor for x-ray imaging.

Another critical requirement may be the overall focal plane size requirement. In applications where overall focal plane size has merit over total resolution, the cost associated with larger CCDs may be prohibitive.

2.1 Fiber Optic Faceplates, Reducers and Magnifiers

The adaptation of fiber optic bundles to the surface of CCDs has greatly increased the number of applications available to CCD imaging. Whereas some instrumentation requirements have necessitated the use of image intensification, losses in signal as the result of lens coupling the CCD to the intensifier can be eliminated. By coupling a specific fiber optic faceplate, reducer or magnifier, the fiber optic output from the intensifier can be coupled directly into the fiber optic input to the CCD, thus reducing signal loss (Figure 1).

Applications requiring the use of x-ray imaging are adapted by the use of phosphors that emit in the usable spectrum of the CCD. It is not technically feasible to deposit these

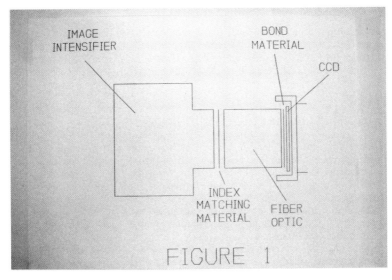

Figure 1

phosphors directly to the surface of the CCD. However, these phosphors can be readily attached to fiber optic faceplates, reducers or magnifiers (Figure 2).

At Xedar, we have a vast experience of coupling these fiber bundles directly to the surface of CCDs. Our capabilities include bonding both large area CCDs and linear single line detectors. Special techniques allow this bonding without significant resolution loss and the ability to withstand temperature ranges of +45° C to -60° C.

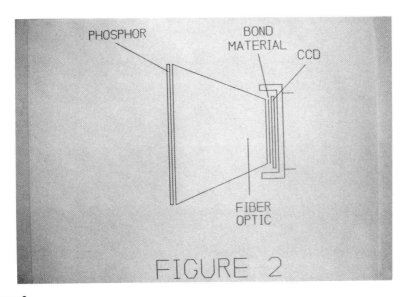

Figure 2

2.2 Multiple Fiber Optics for Large Focal Planes

When the overall area of the focal plane requirement exceeds that which is feasible from a single detector or from a single fiber optic, multiple fiber optic bundles may be butted together to form a single focal plane (Figure 3). Xedar has demonstrated the capability of butting fiber optic bundles in this manner with alignment capabilities to the single CCD pixel.

3 INNOVATIVE ELECTRONIC CONFIGURATIONS

Camera development for scientific applications has many critical aspects. Achieving the highest possible dynamic range in applications where the maximum available signal is quite small requires many special considerations. Removing dark current generation from the CCD by means of cooling has improved performances significantly. Elimination of other noise sources has presented altogether different problems.

Specific techniques developed over years of design work at Xedar have reduced total system noise to a level that is now only limited by the CCD itself. Of these design techniques, the use of opto isolation, specialized correlated double sampling and board layout expertise have given Xedar designs the advantage of extreme noise reduction.

CCDs have been available in large formats (>1K x 1K) for quite some time. These CCDs, with multiple parallel outputs, are capable of operating at very high frame rates. These high speed devices present a significant challenge, as the readout electronics must be multiplied and eventually multiplexed together. Xedar has developed specific techniques in accomplishing these tasks in conjunction with delivering this data to the data acquisition site at the frame rates offered by the CCD.

4 PRODUCTS AND PROGRAMS

In the course of developing application-specific imaging subsystems for the OEM community, Xedar has made available many of the products manufactured for these

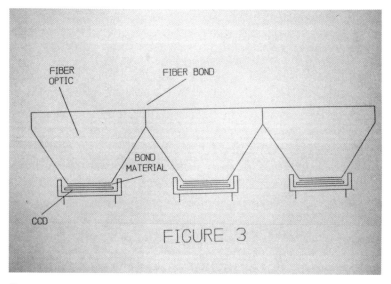

Figure 3

programs. These products can be utilized in their current configurations or modified to suit any number of applications. All cameras may be configured for fiber optic or lens input.

4.1 Full Frame 1024 X 1024 Pixel Camera (Figure 4)

This camera is versatile, high performance, and especially well suited for integration into complete systems It is a cooled, 1024 x 1024 pixel, full field imaging CCD camera, operating at 500 Khz output data rate. The camera head contains the sensor, the clock and drive circuitry, and the analog to digital conversion circuitry. This camera is capable of integration times as long as 10 seconds with 12 bit dynamic range.

Applications range from medical to industrial diagnostics and inspection. With multiple optical input options, this camera lends itself to various x-ray imaging applications, such as non-destructive test and inspection.

The camera head can be configured to custom requirements and combined with Xedar optional accessories, such as computer interface card, power supply and cabling, to provide a complete imaging system.

Features:

KAF-1000, 1024 X 1024 pixel CCD
Adaptable to customer specifications
Software addressable operation
Integration times to 10 seconds
12 bit dynamic range
Parallel differential digital output
Variable gain adjustment
24 μm X 24 μm pixel size
Cooled to -15° C.
Data rate of 500 Khz
RMS noise < 20 electrons
Frame rate 2.3 seconds

Figure 4

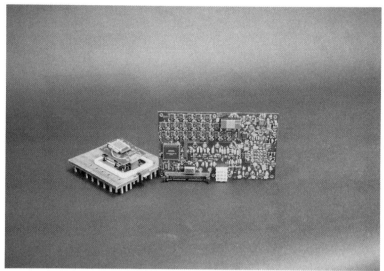

Figure 5

4.2 Full Frame 1280 X 1024 Pixel Board Level Camera (Figure 5)

Based on the Kodak KAF-1300L, this unique board level camera is high performance and simple to use. It is especially well suited for integration into various types of equipment for any number of applications. Its configuration as a board level camera allows convenient adaptation into equipment in which box type cameras are inconvenient to use. At the user's discretion, Xedar can incorporate this camera into a variety of housings, allowing its use as a stand alone camera.

Due in part to its unique configuration and performance level, this camera is ideal for analytical applications such as surface analysis and many forms of spectroscopy.

Features:

KAF-1300L 1280 X 1024 pixel CCD
Software addressable operation
Variable integration time
12 bit dynamic range at room temperature
Parallel differential digital output
16 μm x 16 μm pixel size
Antiblooming protection
2.5 Mhz output pixel rate
RMS noise < 30 electrons

4.3 Full Frame 2048 X 2048 Pixel Camera (Figure 6)

The 2048 X 2048 full frame camera is based on the Thomson THX7899M high performance CCD. The unique architecture of this CCD allows full frame data from one to four outputs. Taking advantage of this configuration, this camera is capable of multiple output schemes. At the output of the camera are two output formats. Parallel differential digital data is available at the camera for applications where the image processing is in close proximity to the camera. For applications where remote operation of the camera is required, a dual fiber optic data link is also available. This fiber optic data link is capable of delivering image data over several hundreds of meters without data loss.

Figure 6

To compensate for dark current generation as well as heat generated by the electronics, the camera is featured with a peltier cooler and liquid heat exchanger. In combination, cooling to -45° C is achieved. All camera operational controls such as 2 X 2 pixel binning, are completely software addressable.

Developed for the National Solar Observatory, Sun Spot New Mexico, this camera lends itself well to various astronomical applications. Versatility of operation makes this camera desirable for scientific and industrial applications as well.

Features:

THX7899M 2048 X 2048 full frame CCD
14um X 14um pixel size
16 Mhz pixel rate output
Cooled to -45 degrees C
12 bit dynamic range
Parallel differential digital output
Fiber optic data link output
2 X 2 pixel binning
Variable integration time
RMS noise <35 electrons
Software addressable controls

4.4 Large Format TDI Scanning Camera (Figure 7)

The large format time delay integration (TDI) camera is the latest development for large area high resolution imaging. This camera contains six Kodak KAI-0360 TDI CCDs, each having 1100 X 330 pixels, optically butted together with the use of tapered fiber optics. Special techniques allow this assembly to have better than a single pixel registration over the entire image area. The seams at the butted points of the fiber optic are assembled in a manner that provides continuous imaging over those areas. Across the imaging area, 6600 pixels by 330 pixels are imaged as a single detector. By providing motion, either movement of the camera or the target medium, the available area scanned in the TDI direction is limitless.

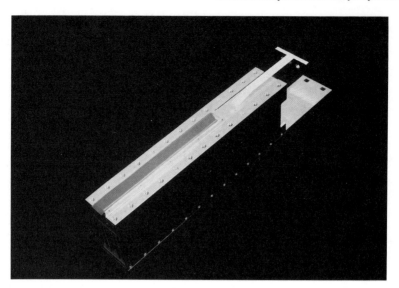

Figure 7

The data from each of the six detectors is read out in parallel. Within the camera, all data is then digitized and serialized, delivering continuous data at 8 million pixels per second. Data is then available from a single fiber optic data link with all control signals software addressable from an RS232 control interface.

The camera requires no cooling other than room temperature air flow for heat evacuation. For the entire camera assembly, RMS noise is specified at < 35 electrons.

Features:

Six Kodak KAI-0360 CCDs
6600 pixels X 330 TDI stages
24 μm X 24 μm pixels
8 Mhz pixel rate
12 bit dynamic range at room temperature
RMS noise < 35 electrons
2 X 2 pixels binning capability
RS-232 data control interface
Single fiber optic data output
Compact size

5 CONCLUSION

As advances in technology continue to produce more sophisticated detection devices, Xedar Corporation is committed to the development of new methods of operation. Both from a mechanical and electronic perspective, Xedar's charter is to provide the most advanced and cost effective ways to integrate these new developments into problem solving tools. With these advances in detection and operation, more and more applications are now finding solutions in imaging. Whether in science or industry, Xedar's ability to provide imaging solutions is without compromise.

A REVIEW OF IMAGING OPTICS FOR THE ULTRAVIOLET

Donald K. Wilson, President
Optics For Research, Inc.
P.O. Box 82
Caldwell, NJ   07006

1 INTRODUCTION

The recent surge in ultraviolet imaging applications, along with availability of UV-enhanced CCD and CID cameras, is caught short by a sparse selection of optical objectives. This suggests new marketing opportunities for designers and manufacturers of such equipment.

2 BACKGROUND

For half a century, infrared technology has dominated in imaging outside the visible because (a) defense spending has concentrated on the infrared in order to detect the presence of IR emitting military engines, (b) IR detection is relatively simple, and (c) the UV has simply not been important in defense applications.

For these reasons, there has been less interest in ultraviolet applications, so the UV field has suffered a lack of funding attention. In recent years, real-time imaging in the UV has been made possible with enhanced CCD and CID cameras.

1.1 Cameras for the UV

"Back-thinned" (UV enhanced) CCD cameras and CID cameras are useful for imaging in the UV. The CID camera, with its inherent high quantum efficiency in the UV, enables us to see real-time images down to at least 180 nm. Between the vacuum and the visible, the UV-enhanced CID camera averages 25-40% quantum efficiency (Figure 1).

1.2 Front-End Optics

For three and a half centuries the laws of reflection and refraction have been known, but there is hardly anything new in optics. Neither microscope objectives, camera lenses, and telescope objectives, nor the methods of manufacturing them, have changed significantly for more than a century.

In the visible spectrum, for example, the newest and most exciting addition to the world of optical components, the diffractive element, was actually achieved by Lord

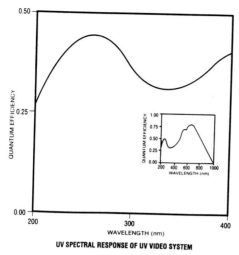

Figure 1 *Spectral response of a UV enhanced CID video camera*

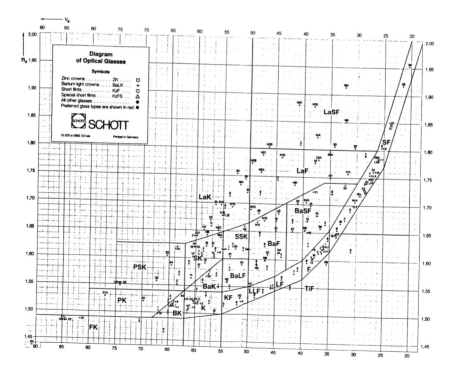

Figure 2 *Refractive indices for selected optical materials*

Rayleigh 124 years ago. Over fifty years ago, diffraction limited photo reconnaissance camera lenses were flying over World War II Europe (Kodak Aero Ektar series).

The only really new development in optics is thin film technology, which has brought us antireflection and dielectric coatings. Also, high speed and "user friendly" computer programs allow rapid design of optical systems.

1.3 Material Choices for the Designer

For visible applications, the designer can choose from more than a hundred glasses and crystals covering an enormous range of refractive indices. For the infrared, there are fewer material choices, but an adequate selection of indices is available (Figure 2). However, for UV, material choices for the designer are few.

1.3.1 Fused Silica. This is the workhorse of all UV optics, readily available from at least a dozen reliable manufacturers worldwide. Fused silica, or fused quartz, is manufactured in a wide variety of grades, with a minimum cost choice according to the application. Fused silica is usable down to 180 nm, and is the least expensive of the UV transmitting materials.

Of all optical materials, fused silica is the toughest. It is extremely hard, has a very low thermal expansion index, and is easily worked in the optics shop.

1.3.2 Calcium Fluoride. After fused silica, CaF2 is the most common and least expensive of the UV materials. It is transparent down to around 120 nm. However, CaF2 is somewhat difficult to grind and polish, and achieving a very smooth, sleek-free surface requires the skills of an experienced optician.

CaF2 is soft, cleaves easily, and is sensitive to rapid temperature change. What restricts the designer are its low refractive indices.

1.3.3 Lithium Fluoride. LiF transmits slightly lower than CaF2, around 112 nm, and is more difficult to work optically. Achieving a sleek-free and pit-free surface requires the skills of an experienced optician, and even then, a little bit of good luck helps.

LiF is more sensitive to thermal changes than CaF2, and is soft and fractures easily. Its indices of refraction are close to those of CaF2, but it is more expensive.

1.3.4 Sapphire. Because sapphires are second to diamonds in hardness, sapphire lens production can be extraordinarily expensive. Although high indices offer the designer a wider latitude of choices, sapphire lenses are rarely used in UV systems.

In addition, its birefringence creates problems in some applications, for example if a prism is required.

1.3.5 Other Materials. There are some glasses that are usable down to around 360 nm, but no further. Other halides (fluorides, chlorides, bromides, iodides) transmit in the UV, but factors of cost, hygroscopicity, and sensitivity exclude them from the small group of acceptable materials.

The result of this lack of materials is that UV lens systems can be narrowly limited. For example, UV achromatic lenses are usually of low numerical aperture (high f/#), narrow field of view, etc. when compared to visible spectrum counterparts. In fact, it is rare that more than two materials (fused silica and CaF2) are used.

2 TYPES OF OBJECTIVES AND THEIR APPLICATIONS

2.1 Reflective Systems

Reflective systems have two major advantages: freedom from chromatic aberration, and assuming all-spherical surfaces,low production cost (Figure 3).

Catadioptric systems, which combine refractive and reflective elements, can be more flexible than all-reflective, but require correction of chromatic aberrations. The major disadvantage of reflective systems is that they will usually have an obscuration.

The Schwartzschild microscope objective is all-reflective. It has no chromatic aberration, an advantage, but it has a central obscuration, which in some applications is intolerable (Figure 4).

All-reflective and catadioptric objectives most commonly appear as microscope objectives (Schwartzschild design) and telephoto lenses.

2.1.1 Long Distance Microscope Objectives. These are used in applications that require inspection of a small object at a remote location, such as in-process inspection in manufacturing, or forensic or biological applications that detect and study luminescence.

Such objectives utilize a modified telephoto objective, refractive or catadioptric, and look at areas a few millimeters in diameter and meters in distances. The image of the object fills the entire detector array area in the camera, allowing high magnification of very small details.

2.1.2 Telephoto Lenses. Looking at a distant object with telescopic magnification requires an optical system that will fill the camera detector array with a very small field-of-view object. A classical telephoto lens, basically nothing more than a telescope with a camera attached, is usually catadioptric.

Observing and recording the UV emission from a rocket engine is a natural application for a telephoto lens on a UV-enhanced CCD or CID camera. Whereas the IR characteristics of rocket engine plumes are well documented, little, if anything, is known about their UV emissions. The analysis in real-time of the UV fingerprint of the plume is possible with the use of CID cameras.

2.2 Refractive Systems

Refractive systems have the advantage of being on-axis and, therefore, are directly similar to visible counterparts.

2.2.1 Microscope Objectives. Classical microscopy in the UV has been practiced for quite some time. An adequate number of UV microscope objectives with magnifications from 2X to 40X are currently manufactured. With the UV-enhanced CCD and CID cameras, and a wider choice of objectives, new applications in UV microscopy are viable.

There is a need for a UV "microscope" objective with short focal length and long working distance, for use as the focusing component for high energy, UV excimer lasers in micromachining, microlithography, materials processing, and biological studies.

The objective should focus the excimer laser beam to micron size spots, and have sufficient resolution and transmittance in the visible for white light monitoring, when desired, or direct imaging in the UV by using a UV transmitting, visible blocking filter.

A long working distance objective avoids waste vapor deposits on the first surface.

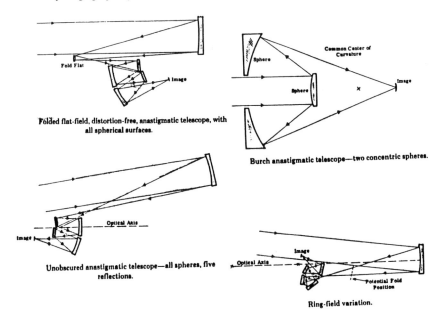

Figure 3 *Reflective optical systems*

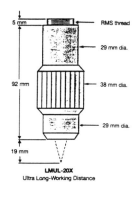

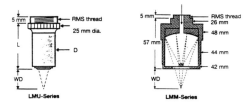

Figure 4 *Refractive optical systems*

Infinity-conjugate, UV achromatic focusing objectives, capable of handling high optical powers, are available. Objectives with magnifications from 2X to 40X, and working distances from 2 mm to 19 mm are also commercially available.

2.2.2 Standard "Camera" Lenses. "Ordinary" picture-taking of a scene uses a lens very similar to a standard camera lens. Biological studies of an area, e.g. a tropical rain forest, in the UV with the visible blocked, will reveal the dependence of natural systems upon the UV.

UV camera lenses are all-refractive. However, because of the limited choice of materials, these lenses are "slow" - they have a small aperture-to-focal length ratio (high f/#). Therefore, use in low light level applications is restricted.

3 AN ESPECIALLY INTERESTING APPLICATION

Since the revelation of atmospheric ozone depletion, biologists worldwide have been concerned about the effects of increased UV radiation on the world's food chains. Life on our planet has evolved with energy input from the sun, and life forms obviously respond to more than just the visible spectrum. What happens in the biosphere when radiation characteristics change?

It is known that atmospheric UV penetrates up to 30 meters into tropical waters. Many life forms in the underwater world use the UV for basic vision, identification of prey and predator, reproduction, and defense.

Since the oceanic food chain begins on the microscopic scale, working up the line from plankton to whales, what happens to the food chain in the planet's oceans as UV radiation increases?

The University of Hawaii Institute of Marine Biology proposes to do video recording in the close sub-surface environment of tropical reefs. Being able to see life activities underwater in the UV, in real time, with the visible blocked, will generate much new information.

4 SUMMARY

New interest in UV imaging is now emerging because of improvements in detection equipment. Whereas UV objectives ought to be readily manufactured, the historic lack of applications in this field has created a dearth of off-the-shelf optics.

Applications in industrial processing, biological and medical sciences, forensics, chemical analysis, and defense can now be undertaken in the UV. All this represents a marketing opportunity, and fertile ground for technological advances.

References

1. D.M. Aikens, 'Beginning the Design of a UV Microscope Objective', formerly with KLA Intruments, Inc., CA.
2. CID Technologies, Inc., 'SICAM' brochure, NY.
3. Schott Optical Glass, Inc., PA.
4. D.R. Shafer, *Applied Optics*, 1978, **17**, 1072.

5. Optics For Research, 'MicroSpot Focusing Objectives' catalog, NJ, pp. 4, 7.
6. Dr. G. Losey, 'Ultraviolet Imagery', Hawaiian Institute of Marine Biology Technical Report #41, 1995.

ECHELLE SPECTROSCOPY AND CCDS, AN IDEAL UNION FOR FIBER OPTIC RAMAN SYSTEMS

Michael M. Carrabba, Job M. Bello, Kevin M. Spencer and John W. Haas, III

EIC Laboratories, Inc.
111 Downey Street
Norwood, MA 02062

1 INTRODUCTION

In its simplest form, the Raman spectroscopic analysis consists of "exciting" a sample with monochromatic laser light, returning the scattered light to a spectrograph, and measuring the spectrum with an appropriate light detection device at the spectrograph's focal plane. Typically in unenhanced Raman spectroscopy, about $10^{-3}\%$ of the intensity of visible light impinging on a molecule will be scattered, mostly elastically at the excitation frequency (Rayleigh scattering). However, about 1% of the scattered light will be at frequencies corresponding to combinations of the exciting light and the molecular vibrational frequencies (the Raman effect). The spectral resolution of the scattered light will yield a series of sharp lines which, like the infrared spectrum, give information about bonding and structure. Furthermore, the sharpness of the lines makes possible the identification of several species simultaneously.

In the past, Raman spectroscopy has been underplayed in chemical analysis due to the size and complexity of the laser and spectrograph previously considered necessary to obtain sufficient excitation intensity, reproducible sampling geometry, spectral resolution, and separation of the weak Raman spectrum from the intense Rayleigh line. However, recent advances in Rayleigh line filtering[1] and in fiber optic sampling,[2-4] combined with advances in solid state laser technology and charge coupled device (CCD) detectors support the wider application of the technique in a variety of new arenas.

Most spectrographs suffer from the fact that they can either obtain the complete Raman range (0-4000 cm^{-1}) with limited resolution or high resolution with a limited range. The limitations are usually imposed by the detector element, slit size, optical components, and the flat field focal plane of the spectrograph. For example, a 4000 cm^{-1} spread across a one inch focal plane with 1000 elements in the spectral dimension would yield a minimum resolution of ~12 cm^{-1}(3 pixels), assuming perfect imaging of the entrance slit.

One method which has been previously discussed for both Raman[5] and emission spectroscopy is to use an echelle spectrograph with a two dimensional CCD detector.[6-10] An echelle spectrograph utilizes a high efficiency, coarse (30-100 lines/mm) diffraction grating in relatively high orders and large angles of incidence. When used in this configuration, the multiple diffracted orders are spatially overlapping. Thus, a secondary dispersive element is used in an orthogonal direction to separate the orders. This enables both high resolution and a large spectral range in a package which can occupy less than 10% of the volume of a conventional high resolution Raman instrument.

This paper will discuss the development and field testing of a ruggedized fiber optic echelle Raman spectrograph. The echelle spectrograph is designed to have both a high light collection efficiency (f/2.8) and a small size (focal length 150 mm), without sacrificing resolution, an impossible feat for traditional spectrographs with one dimensional linear dispersion. The lens based optics are matched to a fiber optic input to provide 1:1 imaging at the focal plane. The spectrograph employs no slits which, combined with the shorter light paths and fewer optical surfaces, makes the optical throughput of the instrument higher than traditional high resolution Raman spectrographs.

2 EXPERIMENTAL

Two echelle spectrograph designs are illustrated in Figure 1. To optimize performance, separate layouts for visible (400-800 nm) and near infrared (NIR) (600-1100 nm) operation were required. The main differences were the focal lengths and f/no of the imaging optics and cross dispersion prism element(s). The NIR version utilizes a combination of a 150 mm focal length (f/2.0) collimating lens and a 150 mm focal length (f/2.5) focusing lens. Both lenses were designed and fabricated to match and optimize the optical throughput of the spectrograph and to minimize optical aberrations over the complete range. The throughput of each lens in the region 600-1100 nm is ~96%. The visible version employs a combination of a 180 mm focal length (f/2.8) collimating lens and a 180 mm focal length (f/2.8) focusing lens. Both lenses were obtained from Nikon.

The NIR version required two SF-10 prisms to completely separate the orders. The visible version employs a single SF-10 prism. Both systems incorporated a commercially available 52.65 g/mm echelle grating (Milton Roy 35-15-415), which was operated in the Littrow configuration at its blaze angle of 63.5°. The angle of the incident ray above the meridian plane was 9°.

The imaging system was an SDS9000 Detection System with a forced air thermoelectrically cooled CH250 camera head supplied by Photometrics Ltd. The spectra presented in flat field at the output of the collimating lens of the spectrograph were imaged onto the EEV CCD15-11 charge coupled device (CCD) detector array (27.6 x 6.9 mm). The array had 1024 x 256 pixels (27x27 μm) and had MPP capabilities.

3 RESULTS AND DISCUSSION

The objective of our efforts was to fabricate a rugged echelle spectrograph which can achieve a high degree of dispersion in a very compact size (600 x 300 cm). The spectrograph needs to obtain the entire spectral range with a resolution of ~1 cm^{-1} without moving or shifting the echelle grating or any other optical components. A predispersion element is necessary to separate the orders so that they will fill the vertical height of the detector. Optical analysis of a variety of echelle designs conducted in this laboratory has resulted in this novel two prism predisperser design (NIR version) in order to minimize two nonlinearities common to this type of spectrograph: 1) curvature or "bowing" of the individual orders, and 2) tilt or "twist" of the spectral lines. These nonlinearities affect the

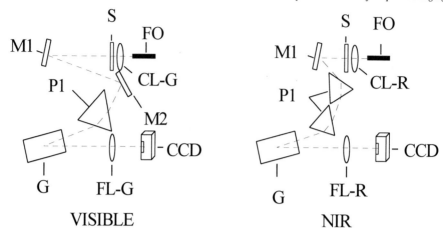

Figure 1   *Refractive Littrow echelle spectrograph for Raman spectroscopy. Left:*
Visible version (400-800 nm). Right: NIR version (600-1100 nm); CL-G -
180 mm f/2.8 Lens, FL-G - 180 mm f/2.8 , CL-R - 150 mm f/2 Lens, FL-R
- 150 mm f/2.8 Lens, FO-Fiber Optic Input, G - 52.65 gr/mm Grating, P-
SF10 Prism, S-Stray Light Aperture Mirror

readout protocol on the CCD detector, which is done most efficiently by first "binning" or summing the pixels together along the height of each order. We have been able to predict and minimize these and other optical aberrations and stray light, since the designs in Figure 1 have also included a full ray tracing analysis for the Stokes spectrum out to 4000 cm^{-1} at several of the common laser excitation frequencies. The use of a grating as the predispersion element was excluded due to stray light concerns.

Table 1 lists the design parameters for the NIR echelle spectrograph, while Table 2 lists the possible grating orders that can be used with various laser excitation sources. The diffracted orders incorporating the desired energy range were calculated using the grating equation:[11]

$$m \; \lambda = a \, (\sin \alpha \pm \sin \beta) \tag{1}$$

Here, m is the diffraction order, a is the groove separation, α is the angle of incidence with respect to the grating normal, and β is the angle of diffraction. For the Littrow configuration and the 52.65 gr/mm echelle grating, the Raman region of 4000 cm^{-1} is divided into 14 intervals or orders. The free spectral range (F) or the useable wavenumber in each order is given by:

$$m = \nu \, / \, F \tag{2}$$

The calculated spectral range of our system is 295 cm^{-1}/order. The length of each free spectral range on the CCD detector is dependent on the wavelength and the focal length of the system. For the NIR system operating with a 785 nm laser source, the length of the orders range from 15.7 mm at order 42 to 20.8 mm for order 31. The image that is

Table 1 *Design Criteria of the NIR Echelle Spectrograph*

Parameters	Design Criteria
Orders	34 to 56
Excitation Source	632, 647, 752, 785 nm
Spectral Range	600-1110 nm
Raman Operational Range	0-3300 cm^{-1}
Input Aperture	50-200 μm Fiber Optic
Spectral Resolution	~ 1 cm^{-1} with 100 μm Fiber Optic
Focal Plane	Within 25 x 6 mm
Acceptance NA	0.2
Imaging NA	0.2

Table 2   *Diffracted Orders Corresponding to the Raman Energy Interval Out to 4000 cm^{-1} from the Laser Excitation*

Laser Wavelength (nm)	Spectral Range (nm)	Orders
632	632-854	55-41
647	647-873	53-39
752	752-1075	45-31
785	785-1144	43-29

presented to the focal plane is shown in Figure 2 for the white light transmission spectrum for the visible and NIR echelles. The spectra and images were collected using a calibrated lamp source.

As is evident from the image, the spectrum continues beyond the free spectral range in each order at both the high and low frequency ends, offering redundancy with the next higher and next lower orders. However, the intensity is highest in the center of each order so that the signal strength and fraction of the available light dispersed at these redundant frequencies will be low. The spectra shown were linearized, calibrated and spliced by a custom software package. The individual free spectral ranges were first linearized to compensate for the inherent variation of wavenumber per order and then calibrated versus spectral calibration lines. After calibration and linearization, the various orders were spliced together to present the spectrum. Also apparent in Figure 2 is the drop off in the CCD quantum efficiency in the 900-1100 nm region.

The resolution of the system is dependent on both the system design and on the width of the CCD pixel elements. Based on pixels only, the detector limited resolution would be 0.5 to 0.37 cm^{-1} for orders 42 and 31, respectively. The collimating and imaging optics are designed so that the resolution is dependent only on input fiber optic diameter and the natural linewidth of the Raman scattering from the sample. Figure 3 shows the neon emission spectrum obtained from the system and illustrates the resolution capabilities of the echelle. The image and resulting spectrum at the bottom of Figure 3 indicates that the resolution of the system is operating within the design criteria outlined in Table 1.

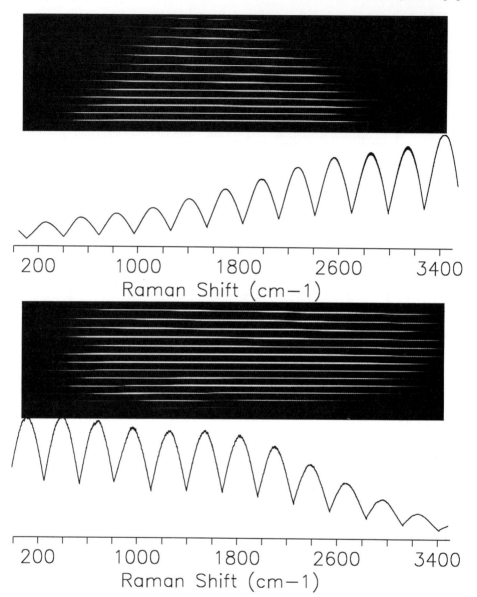

Figure 2   *Top - White light CCD image and spectrum of echelle orders as they are
projected onto the visible echelle spectrograph focal plane for the spectral
range 500-700 nm;   Bottom - NIR echelle spectrograph CCD image and
spectrum for the spectral range 750-1100 nm.*

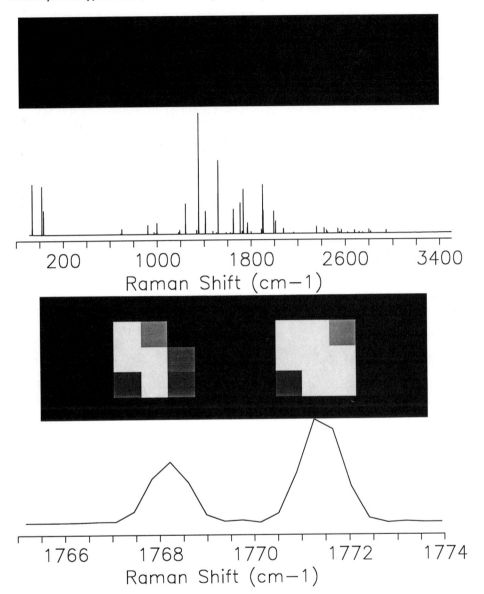

Figure 3 *Top - CCD image and spectrum of the neon emission spectrum (750-1100 nm) for the NIR echelle; Bottom - CCD image and spectrum of two bands separated by 3.1 cm^{-1}. The input was a 50 μm fiber optic input with no slit.*

To illustrate further the resolution capabilities of the echelle, a Raman spectrum and the CCD image of pinene obtained with the NIR echelle system is shown in Figure 4. Pinene was chosen because it has a large number of closely spaced Raman bands. The spectrum was collected with a 1 second acquisition time using 80 mW of 752.4 nm light generated from a Krypton ion laser. The spectrum was corrected for the instrument response function by dividing by the white light spectrum shown in Figure 2.

A field demonstration of the fiber optic echelle spectrograph was recently conducted at a U.S. Army site for the emergency analysis of several suspected Chemical Agent Identification Set (CAIS) ampules. CAIS samples have been previously determined to contain either industrial or agent/chloroform solutions and were used as training sets during 1940-1950's. The suspected CAIS ampules were turned in to a munitions amnesty container at an Army base and had no markings on them to identify their liquid contents. The Army needed to accurately determine the contents of the ampules prior to selecting a disposal method, since transportation off-site is prohibited by law if they contain chemical agents. In addition, this base was slated for closure and could not be shut down until the ampules were properly disposed of. If the ampules contained agent compounds then it could significantly delay the closure process.

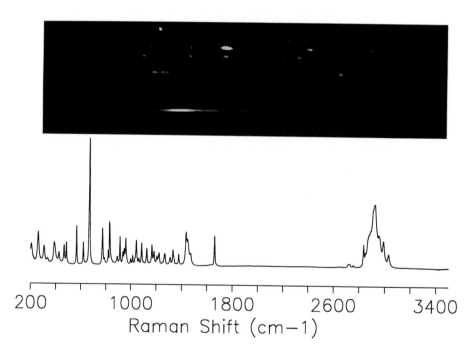

Figure 4   *Fiber optic NIR echelle Raman spectrum of Pinene. Laser wavelength was 752.4 nm with a laser power of 80 mW. Acquisition time was 1 second with a 200 μm fiber optic input.*

A visible ruggedized fiber optic echelle Raman spectrograph was transported via a pickup truck to the remote site. Upon arrival a performance and calibration check was conducted with laboratory prepared samples. The performance checks indicated that the unit required no alignment or adjustment after transportation. The ampules were located inside a secured area which was accessed by deploying a fiber optic cable though a fence. This allowed for the Raman operators not to be required to suit up in heavy protective gear. Ambient temperature was ~30°C. A fiber optic Raman probe[1] and ampule holder were deployed just outside of the munitions bunker in which the ampules were stored.

Figure 6 shows the Raman spectrum of one of the unknown ampules. The sample was a light straw color. The spectrum was obtained in 10 seconds with 53 meters of fiber optic cable. Upon comparing the spectrum with previously recorded standards, it was determined that the sample contained the industrial compound chloropicrin. The results of the field analysis indicated that all of the ampules were industrial compounds and standard chemical disposal method could be utilized.

4 CONCLUSIONS

This paper discussed the development and field testing of a fiber optic echelle Raman spectrograph. The echelle spectrograph was designed so that it could efficiently obtain the complete Raman spectrum with high resolution in a relatively compact size. Field and laboratory results have demonstrated its ruggedness and efficiency. It should be noted that an alternative approach for range and resolution would be to utilize a Fourier Transform (FT) Raman unit based on an interferometer. The FT-Raman technique has been developed primarily for operation at 1064 nm (Nd:YAG laser). However, it has been

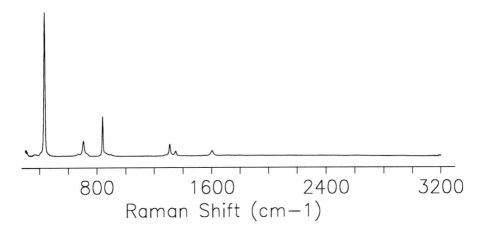

Figure 5 *Fiber optic Visible echelle Raman spectrum obtained in the field of an unknown CAIS sample. Laser wavelength was 514.5 nm with a laser power of 80 mW. Acquisition time was 10 seconds with a 200 μm fiber optic input.*

shown experimentally that the sensitivity of a dispersive spectrometer is much higher than the FT instrument when operating in the visible out to the CCD detection limit (~1100 nm for red enhanced CCD chips), due in part to the extremely high CCD quantum efficiencies and low noise.[12] In addition, field applications of FT-Raman may be limited by the delicate nature of existing instrumentation.

References

1. M. Carrabba, K. Spencer, C. Rich and R. Rauh, *Appl. Spectrosc.*, 1990, **44**, 1558.
2. M. Carrabba and R. D. Rauh, "Apparatus for Measuring Raman Spectra Over Optical Fibers", U.S. Patent 5, 112, 127, 1992.
3. R. Benner and R. Chang, "Utilization of Optical Fibers in Remote Inelastic Light Scattering Probes", Proceedings of the Conference on Fiber Optics. Advances In Research and Development, Kingston, R.I., June 19-23, 1978, 625.
4. M. L. Myrick and S. M. Angel, *Appl. Spectrosc.*, 1990, **44**, 565.
5. M. Pelletier, *Appl. Spectrosc.*, 1990, **44**, 1699.
6. R. Bilhorn, P. Epperson, J. Sweedler, and M. B. Denton, *Appl. Spectrosc.*, 1987, **41**, 1125.
7. P. Epperson, J. Sweedler, R. Bilhorn, G. Sims, and M. B. Denton, *Anal. Chem.*, 1988, **60**, 327A.
8. J. Sweedler, R. Jalkain, G. Sims and M. B. Denton, *Appl. Spectrosc.*, 1990, **44**, 14.
9. A. Scheeline, C. Bye, D. Miller, S. Rynder, and R. Owens, Jr., *Appl. Spectrosc.*, 1991, **45**, 334.
10. C. Bye and A. Scheeline., *Appl. Spectrosc.*, 1993, **47**, 2031.
11. M. Hutley, 'Diffraction Gratings', Academic Press, London, 1982.
12. Y. Wang and R. McCreery, *Anal. Chem.*, 1989, **61**, 2647.

AN APPLICATION OF RAMAN SPECTROSCOPY IN THE PETROLEUM INDUSTRY

Ph. Marteau[1], A. Aoufi[1] and N. Zanier[2]

1. Laboratoire d'Ingénierie des Matériaux et des Hautes Pressions. CNRS.
   Institut Galilée, Av. J.B. Clément, 93430 Villetaneuse, France.

2. Institut Français du Pétrole. 1 et 4 Av. de Bois Préau, 92500 Rueil-Malmaison, France.

1 INTRODUCTION

To improve daily operations, operators of petroleum refineries are showing a strong interest in on-line analysis. Recent advances in optical fibres, chemometrics and instrumentation have made near infrared spectroscopy an attractive analysis tool for the refining process[1,2] but Raman spectroscopy should be also considered[3] since it offers some advantages over near infrared spectroscopy: (i) the whole spectrum extends over the visible region where the transmission of optical fibres is the best; the distance between the sensors located on the unit and the laser and spectrometer can then be as long as some hundred meters, and (ii) the spectrometer can be equipped with a charge coupled device (CCD) detector, which provides simultaneous recordings of different sampling points.

This paper describes the Raman tool developed to control a para-xylene separation process based upon a simulated moving bed chromatography,[4] called ELUXYL (developed by the French Institute of Petroleum, Paris, France). In order to anticipate feed quality changes and optimize process units on the basis of a continuous monitoring of stream qualities, the goal was to obtain the concentration values of the five components at four points of the unit every 15 s. The system has been successfully used since December 1993, working 24 hours a day to control a micropilot unit, and since January 1995 at the Chevron refinery in Pascagoula, Mississippi.

Since the length of this paper has been voluntarily limited, the interested reader will find more detailed descriptions of one or another part of the device in previous publications.[5-8]

2 OPTICAL CONFIGURATION

A schematic diagram of the optical device is shown in Figure 1. The laser beam (514.53 nm of an argon laser) is split towards four optical fibres connected with the four sensors located on the separation unit. The optical fibre length is presently 90 m but can be as long as 250 m, as its transmission is still 50% at this wavelength. The Raman signals are sent back to the spectrometer by four different optical fibres. The four Raman spectra are then simultaneously displayed on four separated areas of the CCD detector. After reading the latter, the spectra are analysed through "SEPAROM" software and concentration profiles are displayed on a second computer using "VISAROM" software.

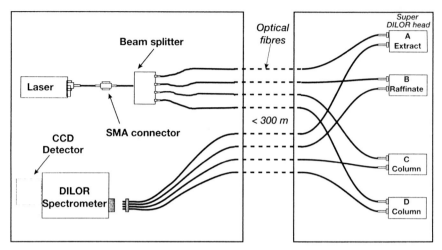

Figure 1 *Schematic representation of the Raman device developed for the control of the ELUXYL process*

2.1 The Super Head Dilor

The optical layout of the Super Head Dilor (SHD), used as sensor, is shown in Figure 2. The SHD is one of the most important parts of the system. It allows one to move the entrance optics far away from the spectrometer. Excitation of the sample and recovery of the back scattered Raman signal are made from the condenser through the window of an optical cell installed on the tubing. The great advantage of the SHD is to bring the Raman signal of the sample onto the collecting fibre without any trace of excitation light which could be able to generate a measurable Raman signal from the silica in the collecting fibre. For that purpose the SHD is equipped with two filters: (i) a narrow band pass interferential filter centered at the laser wavelength which rejects the strong Raman signal induced in the excitation fibre, and (ii) a low pass holographic filter which transmits 80% of the back scattered Raman signal and works as a mirror for the excitation light. The main advantage

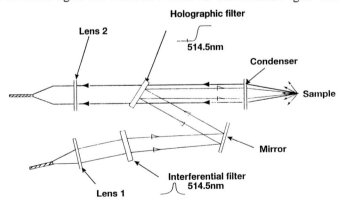

Figure 2 *Optical layout of the Super Head Dilor*

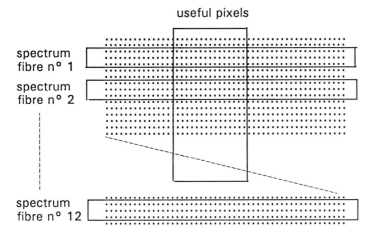

Figure 3 *Up to 12 spectra can be recorded at once on the CCD detector. In fact, the reading is limited to the useful pixels, as simulated by the insert.*

of the SHD is that it solves completely the problem of superposition of the Raman spectrum of the silica on that of the sample, as often occurs.

2.2 Multipoint Analysis.

Up to 12 fibres can be simultaneously connected to the spectrometer. All of them have been carefully aligned in front of its optical entrance so that the corresponding spectra are displayed horizontally on separated areas of the CCD detector. Each spectrum lies on 10 lines of pixels and is separated from its neighbours by the same number of lines.

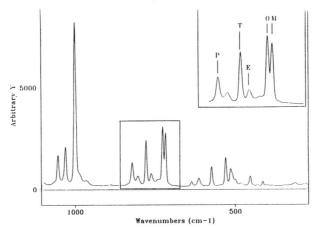

Figure 4 *Typical Raman spectrum of a synthetic mixture of a C$_8$ aromatic cut in toluene (o-Xylene 20.72%; m-Xylene 27.40%; p-Xylene 11.66%; ethylbenzene 17.19%; toluene 23.03%)*

As shown in Figure 3, the reading of the CCD is limited to the rows and columns which are useful for analysis. By so doing one restricts computation time and data volume. As a matter of fact, although the Raman spectrum of the aromatic cut involved in the process extends on a large frequency range, as shown in Figure 4, only a small part of it, shown in the insert, is sufficient to allow determination of the concentrations. There are several advantages in selecting the smallest frequency range including at least one Raman line for each component: (i) the whole spectrum can be recorded without rotating the grating, (ii) the probability for the appearance of spikes is considerably reduced, and (iii) the baseline is more easily recovered even in the case of a fluorescent background.

3 QUANTIFICATION

The principle of the quantification is based upon the linear relationship which exits between the Raman intensity and the molar concentration. The relative concentration of a molecule j in a mixture including n types of molecules is then obtained by calculating the following:

$$C(j) = \left[P(j)/\sigma(j)\right] / \left[\sum_n P(i)/\sigma(i)\right] \tag{1}$$

where $P(j)$ is the integrated (or partial integrated) intensity of the molecule j and $\sigma(j)$ its relative cross-section.

Because of the overlap that always exists between the lines, the $P(j)$ values must be deduced from the experimentally measured integrated intensities $M(j)$ by solving the following system:

$$a_{11} P(1) + ... + a_{i1} P(i) + ... + a_{61} P(6) = M(1)$$

$$...$$

$$a_{1j} P(1) + ... + a_{ij} P(i) + ... + a_{6j} P(6) = M(j) \tag{2}$$

$$...$$

$$a_{16} P(1) + ... + a_{i6} P(i) + ... + a_{66} P(6) = M(6)$$

where the a_{ij} coefficients represents the contribution of the ith molecule on the integrated frequency domain corresponding to the jth molecule. The a_{ij} coefficients are deduced from the Raman spectra of the pure components as being the ratio between the integrated intensities in the frequency domains j and i, respectively.

Finally, the above system can also written as the following:

$$[K] [P] = [M] \rightarrow [P] = [K]^{-1} [M] \tag{3}$$

3.1 Calibration Versus Temperature

In the case of ELUXYL, as for most of the applications, the temperature of the mixture may change from 20 to 180°C, depending on the location of the sensor as well as on the operating parameters. As a consequence of the resulting shifts and broadening of

Table 1 *Influence of the temperature introduced in the program SEPAROM. The true temperature is 21°C. A strong deviation of the results is observed when the calculation is made upon the basis of a temperature of 180°C.*

Compound	True value	$\Theta = 21°C$	$\Theta = 180°C$
Toluene	18.1	18.2	21.6
m-Xylene	21.2	20.9	18.5
p-Xylene	25.5	25.8	28.1
o-Xylene	22.9	22.6	22.5
Ethyl benzene	12.3	12.8	14.1
para-Diethylbenzene	0.0	-0.3	-4.8

Table 2 *Efficiency of the SEPAROM software regarding the influence of the temperature*

Compound	True value	Temperature (°C)				
		21	60	100	140	180
Toluene	18.1	18.2	18.3	18.2	18.2	18.4
m-Xylene	21.2	20.9	21.0	20.7	20.6	20.6
p-Xylene	25.5	25.8	25.6	25.7	25.7	25.8
o-Xylene	22.9	22.6	22.4	22.5	22.5	22.4
Ethylbenzene	12.3	12.8	12.9	12.9	13.1	13.1
para Diethylbenzene	0.0	-0.3	-0.1	-0.0	-0.1	-0.2

the Raman lines, the k_{ij} coefficients of the K^{-1} matrix had to be expressed as a function of temperature. Such a calibration is justified by the results reported in Table 1. A synthetic mixture of known concentrations has been analysed by recording its Raman spectrum at room temperature. If the right temperature is introduced in the computer, Column 3 of Table 1, the obtained concentration values are very close to the right ones whilst strong distortions are observed in Column 4 when the calculation has been made upon the basis of a temperature of 180°C.

The efficiency of the calibration versus temperature is illustrated by the results reported in Table 2. The same mixture as above has been analysed by Raman spectroscopy at various temperatures and now the right temperatures have been introduced in the computer. A good agreement is observed between measured values and right ones with an uncertainty of about 0.5% over the whole 0-100% range.

4 TYPICAL RESULTS

This Raman system was first tested and used for several months to improve the performance of the ELUXYL separation process at the industrial development center of the French Institute of Petroleum. It was then installed on a para-xylene separation unit at the refinery of CHEVRON (USA), where it has been working successfully 24 hours a day

Table 3 *Comparison between Raman and chomatographic (GC) analyses of a raffinate*

	Toluene	m-Xylene	p-Xylene	o-Xylene	Ethylbenzene
Raman	41.42	43.15	2.26	2.56	10.61
G.C.	41.55	42.94	2.07	2.76	10.68

since January 1995. The fibre length between the unit and the control room is 90 m. A laser power of 400 mW is required in order to get an excitation power of 40 mW at the exit of each of the four sensors located on the unit. A relatively small part of the total loss (60%) is due to optical fibre attenuation (23%) and the remainder is due to reflection on optical components (lenses, filters, *etc.*). A recording time of 10 s is enough to get a convenient signal/noise ratio. An example of the four spectra obtained is given in Figure 5, which also shows the spectrum of a neon source taken as a frequency pixel reference. Although these spectra exhibit a rather strong background, due to the fluorescence of impurities contained in the fluid, the results of the Raman analysis could be favourably compared to those simultaneously obtained by chromatography. The results of one of these tests are reported in Table 3.

The concentrations deduced from the Raman spectra are transferred to a second computer where the concentration profiles are displayed. A concentration profile versus the 24 beds of the chromatographic column is represented in Figure 6. The four Raman spectra are recorded every 15 s, which gives five points of measurements per bed. Figure 6 represents the superposition of two successive profiles. The operator can immediately deduce the stability of the unit and easily determine the best withdrawal point for the extract. For the illustrated example, the extract is enriched in para-xylene between beds 17 and 19 and is not too contaminated by the other components.

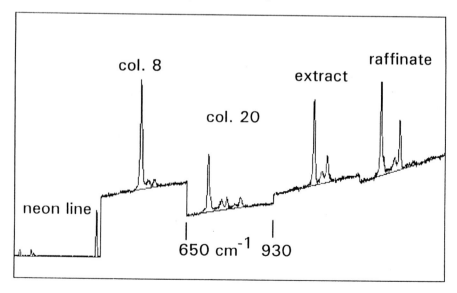

Figure 5 *Simultaneous Raman spectra at different points of the ELUXYL process*

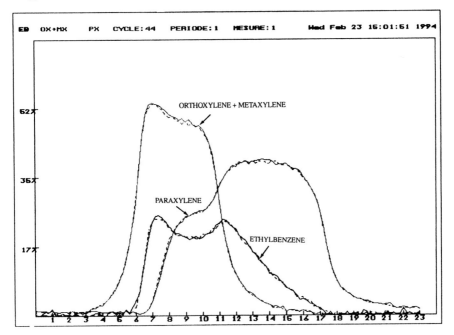

Figure 6 *Typical concentration profiles obtained by Raman analysis on the ELUXYL process*

When the process was controlled by GC, a complete day of stabilization was required before any parameter could be changed. According to the daily operators, 1 h is now sufficient.

5 CONCLUSION

The Raman tool described above has proved its efficiency for two years in real industrial conditions. The operators using it claim it is a powerful tool for monitoring the process. It is noteworthy that the availability of a CCD detector in the optical system has been at the origin of the success. As a matter of fact, it allows one not only to record the whole spectrum at once but also to get several spectra simultaneously, so the on-line remote analysis becomes multipoint.

References

1. A. Espinosa, D. Lambert and M. Valleur, *Hydrocarbon Processing*, 1995, **74**, 86.
2. M. J. Lysaght, J. J. Kelly and J. B. Callis, *Fuel*, 1993, **72**, 623.
3. J. B. Cooper et al., *Appl. Spectros.*, 1995, **49**, 586.
4. G. Hotier and B. Ballanec, *Rev. Inst. Fr. Pet.*, 1991, **46**, 803.
5. Ph. Marteau, G. Hotier, N. Zanier-Szydlowski, A. Aoufi and F. Cansell, *Process Control and Quality*, 1994, **6**, 133.

6. Ph. Marteau, N. Zanier, A. Aoufi, F. Cansell and E. Da Silva, *Analysis Magazine*, 1994, **22**, 32.
7. Ph. Marteau, N. Zanier-Szydlowski, A. Aoufi, G. Hotier and F. Cansell, *Vibrational Spectrosc.*, 1995, **9**, 101.
8. Ph. Marteau and N. Zanier, *Spectroscopy*, 1995, **10**, 26.

DEVELOPMENT OF AN OPTIMIZED CCD ARRAY-DETECTOR, DIODE LASER/FIBER-OPTIC RAMAN SYSTEM FOR FIELD ENVIRONMENTAL ANALYSIS

Daniel A. Gilmore and M. Bonner Denton

Department of Chemistry
University of Arizona
Tucson, AZ 85721

1 INTRODUCTION

As part of its mission to characterize, remediate, and monitor superfund sites after closure, the U.S. Environmental Protection Agency requires field instrumentation to continuously monitor low levels of environmental contaminants.[1] The application of laser Raman spectroscopy to field analysis could provide a highly flexible and selective environmental detection system. The combination of technology that has recently become available (diode lasers, fiber-optics, holographic optics, high throughput imaging spectrographs, and CCD detectors), enables the construction of a portable Raman instrument capable of making a significant impact in the areas of analysis and monitoring under a variety of sampling conditions.

There have been many reports discussing the attributes of fiber-optic Raman,[2-15] diode laser Raman,[16-18] and diode laser/fiber-optic Raman systems.[19-23] Diode lasers allow the instrument to be compact and have low power requirements, enabling portability. The near infrared wavelength (~785 nm) of the diode lasers also reduces interferences from fluorescence. The fiber-optic probe technology provides fast and easy remote sampling with no alignment and accommodates in-situ sampling (down a well, inside tanks and drums, through the side of bottles, etc.).[22] Holographic filters[23-27] allow the use of small, optically fast single-grating spectrometers greatly increasing the light throughput of the system. With the combination of imaging spectrometers and scientific charged coupled device (CCD) cameras, a new level of sensitivity and flexibility can be realized for Raman spectroscopy through multipoint sampling in a single acquisition.[21]

The vibrational spectrum obtained with Raman spectroscopy is complementary to that from FTIR as it depends on polarizability whereas for IR absorption, there needs to be a dipole moment change within the molecule. Therefore, some molecules that have no dipole moment have no IR spectrum. Also, IR spectroscopy becomes very difficult in aqueous solutions due to strong water absorptions and the need for a window material that is not water soluble (normally salt windows are used for IR cells). The Raman spectrum of water does not cause such problems, and no cell is needed with fiber-optic probes. Together, Raman and IR spectra give the complete vibrational spectrum of a molecule.

In contrast, although mass spectroscopy (MS) can have very high sensitivity, it necessitates a much more complicated system and is not generally suitable for field applications. Different samples need special introduction methods along with various types of ionization, thus no one MS system is applicable to all samples. MS systems also need a vacuum system, which makes spectrometer portability a problem.

While the benefits of using diode laser/fiber-optic Raman have been noted, quantitative information concerning detection limits, which are important for real world applications, is lacking. In this work, quantitative detection limits are determined for several compounds. Calibration procedures are developed for the Raman system. Different spectrometers and fiber-optic probes are evaluated. The benefits and limitations of the current instrumentation will be discussed.

2 EXPERIMENTAL

A large part of this project was the development, evaluation, and optimization of the various components of the experimental apparatus. Pursuant to that, details of the important components of the apparatus will be described and evaluated in this section. A general block diagram of the main components of the diode laser/fiber-optic Raman spectrometer (diode laser, optical isolator, fiber-optic probe, spectrograph, CCD detector, and computer) is shown in Figure 1. The components were chosen with portability in mind for use as a field instrument. Properly configured the apparatus would occupy a few cubic feet, weigh less than 100 pounds and require only a few amps at 115 volts.

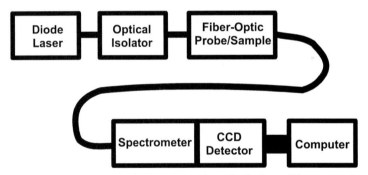

Figure 1 *A general block diagram of a diode laser/fiber-optic Raman spectrometer*

2.1 Diode Lasers

The two different diode lasers used in this work provided a small and efficient Raman excitation source. The diode lasers' wavelengths were tunable (approximately ±5 nm via temperature tuning) in the vicinity of 782 nm. The near infrared wavelength helped alleviate fluorescence, which is often a problem in Raman spectroscopy. The first laser used was a Melles Griot, Model S1386, which uses a Sharp LT025-MD diode and has a maximum rated output of 50 mW. This laser diode could be run at 60 mW but this lead to a shorter lifetime than expected. The expected lifetime of this type of diode laser is 10,000 to 50,000 hours. The diode laser driver (Melles Griot, Model 06DLD203) had both the power supply and the thermoelectric cooler controller (TEC) in a single small (7" x 4" x 12") unit. The drive current for this diode was 160 mA at 2 volts, and the TEC required only a few mA current once the operating temperature (~15° C) was reached. This diode laser gave a very stable output power and showed good frequency stability. There is a potential problem when using this type of diode laser for Raman due to a tendency to change frequency, referred to as "mode hopping". This is a well known phenomenon in laser diodes and can be kept to a minimum if they are operated properly. Proper operation is to turn on the thermoelectric cooler and let the temperature stabilize before the current is applied to the diode laser and keep the current and temperature constant. It is important to know that the wavelength of the diode laser varies by 0.24-0.3 nm/°C if the current to the laser is turned off while the temperature is adjusted. If the laser current is on while the temperature is adjusted, the wavelength tuning is on the order of 0.06 nm/°C until the laser mode "hops" and changes by as much as 1 nm or more. To achieve the best stability, it may be necessary to try different temperatures near the desired wavelength to ensure that it is not operating near an unstable "mode hopping" point. Lasers that are just now becoming available address this stability problem and will be discussed later.

Due to degradation of the output power of the Sharp diode (possibly caused by operating at above rated power), it was replaced with an SDL Inc., Model SDL-5412-H1; the SDL diode has a maximum output of 100 mW with only ~120 mA input current. In addition to having a higher power, this laser diode had a TEC built into the TO-3 package. This had the advantage of the TEC only cooling the laser diode chip and not the entire package, mount, and lens as was the case in the Melles Griot laser. Also, there was no danger of condensing water on the diode package with the SDL laser which was a problem with the Melles Griot laser since often the lasers are run at ~15°C. The SDL laser diode has been very reliable and we have been using the same diode for nearly two years.

Optical feedback to the diode laser such as reflections from the fiber-optics or the sample, caused instabilities in the laser frequency and power. In order to prevent these instabilities, an optical isolator (Optics for Research, Model IO-7-780-LP) was added to the system directly after the diode laser and provided -38.0 dB of isolation. This allowed stable diode laser operation for extended periods of time regardless of the fiber-optic coupling or sampling.

A narrow bandpass filter (Barr Associates, Inc.) was used following the optical isolator to remove non-laser emission from the output. Although the intensity of this emission was orders of magnitude less than the single laser frequency, it overlaps the Raman region and causes interferences. This bandpass filter was necessary only with the first fiber-optic probe used (see below). A diode laser achromatic lens (Melles Griot, Model 06LAI003) was used to launch the laser light into the fiber.

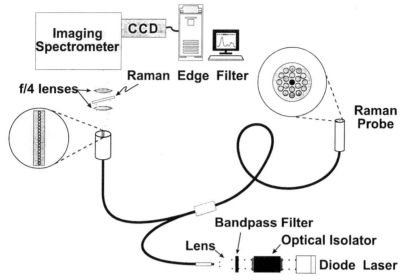

Figure 2 *Schematic diagram of the diode laser/fiber-optic Raman spectrometer*

2.2 Fiber-Optic Probes

Fiber-optic probes enable flexible remote sampling and require no aligning from one sample to another aside from positioning the probe at the sample. Depending on the type of probe used, the sample can be up to a few hundred meters from the Raman system. The first fiber-optic probe used (Chromex, Raman fiber-optic probe) could be extended up to three meters away from the spectrometer. The probe end of the fiber bundle consists of a single laser delivery fiber (the center darkened fiber in the Raman probe, Figure 2) surrounded by 18 collection terminating fibers. The surfaces of the 18 collection fibers are arranged in a line at the spectrometer end of the bundle to better match the slit shape, enabling efficient coupling to the spectrometer. This end of the fiber bundle is fixed in an optimized position for collection and imaging onto the slit of the spectrometer. Two *f*/4 achromatic lenses (Newport, Model PAC052) are used to image the fiber bundle on the slit of the spectrometer (when the Chromex spectrometer was used). A Raman holographic edge filter (Physical Optics Corporation) is placed in the collimated region between the two lenses to block the laser wavelength. This configuration allows rapid remote sampling normally by simply dipping the end of the probe into the sample solution without any of the realignment associated with a normal Raman (non-fiber-optic probe) arrangement.

Though this probe has a very high throughput leading to high sensitivity, there is a problem due to the Raman signal from the collection fibers. This stems from the Rayleigh scattered or reflected light from the sample that is incident on the collection fibers. This light, which is at the frequency of the laser, generates Raman scattering in the collection fibers. Since the interaction length is proportional to the fiber Raman signal, the usable length of this design of probe is limited. Figure 3 shows the Raman spectrum

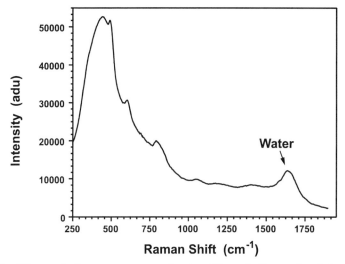

Figure 3 *Fiber-optic Raman spectrum of water. The only Raman band due to water is at 1640 cm⁻¹, the other features in the spectrum are from the silica fibers*

of water which has only one feature in this region at 1640 cm⁻¹, the remaining features are due to the collection fibers.

The second probe used (K-Probe) was built in our laboratory based on a design by Kaiser Optical Systems Inc. This design is shown in Figure 4. This probe head incorporates filtering of the incoming laser light before it strikes the sample and filtering of the Raman scattered light before it enters the return fiber. In this way, only Raman scattering from the sample (no Rayleigh scattering or reflected light) reaches the collection fiber so that no Raman scattering of the fiber is generated allowing long lengths of fiber to be used.

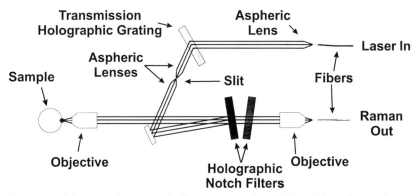

Figure 4 *Schematic diagram of a fiber-optic probe (K-Probe) based on a design by Kaiser Optical Systems Inc. This probe contains filtering to remove laser light that can give rise to Raman signal from the fiber*

The probe head starts with an aspheric lens to collimate the light from the laser delivery fiber. Next a transmission holographic grating and a spatial filter made of two aspheric lenses and a slit allow only the light at the laser excitation frequency (785 nm) to pass. This filters out any extraneous emission from the diode laser and any Raman scattering generated in the delivery fiber. Following the spatial filter, the light is reflected off a mirror to a holographic notch filter that is highly reflective at the laser wavelength but will pass the Raman scattering. The first holographic notch filter reflects all the laser light through the sample objective and onto the sample. The scattered light is collected at 180° and collimated by the objective. The Raman scattering passes through the two holographic notch filters while any Rayleigh scattering (at the laser wavelength) is reflected by the two filters so that only the Raman scattered light is coupled to the return fiber by a second objective lens.

With this probe there is no Raman signal from the fiber present in the spectrum and the probe can be implemented with two optical fibers. This removes fiber length limitation and allows long fiber-optic lengths to be used for remote sensing. Some drawbacks of this design are lower laser power at the sample and less scattered light collected compared to the 18 collection fiber around 1 delivery fiber probe (18-1 probe). The band pass filter used before coupling the laser to the delivery fiber in the 18-1 probe was not required with this probe head design due to the filtering in the probe head. However, the additional optics and filters in this probe head resulted is less laser light reaching the sample. In the case of the 18-1 probe, 75% of the energy measured after the optical isolator reaches the sample (transmission of the bandpass filter was ~81% and the fiber coupling efficiency was ~92%), while with the K-probe head, 47% of the energy reached the sample. For the K-probe, this is lower than could be expected with optimized components that were not available when the probe was built. With the optimized components used in the now commercially available probe from Kaiser Optical Systems, Inc., the throughput should be in the 70%-75% range.

In addition to the loss in laser energy delivered to the sample in the K-probe, more light was collected by the 18-1 probe due to the excitation geometry and the 18 collection fibers. The optical geometry of the K-probe is necessary since the holographic notch filters have an angular dependence and therefore require collimated light. The initial K-probe built used a single 100 µm fiber as the collection fiber. The total signal decreased by a factor of 12.5 with the initial K-probe compared to the 18-1 probe. To improve on the sensitivity of this new probe, a 7 fiber bundle was constructed to be used in place of the single collection fiber in the K-probe. This allowed not only the confocal scattered light to be collected, but also scattered light outside the confocal point in the sample while still maintaining the Rayleigh filtering of the holographic notch filters necessary to prevent Raman signal generation in the fibers. The bundle used 100 µm fibers with 6 fibers around 1 at the probe end and all seven fibers in a single row at the spectrograph end for efficient coupling to the slit. This gave an improvement of a factor of 2.7 times the single fiber yielding an overall decrease of a factor of 4.6 compared to the 18-1 probe. Though this requires additional fibers for the collection bundle compared to the single fiber version, the gain may be worth the added complexity and cost in a given situation.

2.3 Spectrographs

There were two spectrographs used in this study. The first (Chromex, Model 250IS) is a 250-mm focal-length-imaging spectrometer that uses toroidal mirrors to correct astigmatism in the focal plane. This spectrometer, which is $f/4$, was chosen over a higher resolution spectrometer for its high throughput, a necessity for sensitive detection. In addition, it has three gratings mounted on a turret that can be changed under computer control to vary the spectral resolution. This feature is very useful for doing survey spectra followed by higher resolution spectra for quantitative determination. The grating used most often was a 600 grooves/mm (g/mm) blazed at 1 μm and provided broad spectral cover-age of 1675 cm^{-1} in a single exposure. The spectrometer slits were set at 75 or 100 μm giving a resolution of about 10 cm^{-1}. The spectrometer was also equipped with a 1200-g/mm holographic grating blazed at 750 nm for higher resolution when needed. The third grating was 150 g/mm blazed in the visible for use in a fluorescence channel if desired. Also added to the spectrometer was a retro-illumination option which adds the flexibility to easily align various collection optics and fiber-optic probes. A benefit gained by using an imaging spectrometer coupled with the CCD camera as the detector is the possibility for multiple fiber-optic probes to be used simultaneously for multipoint monitoring. This capability was not used in this work.

The second spectrograph used, Kaiser Optical Systems Inc., Model HoloSpec f/1.8, provided a very high throughput ($f/1.8$) imaging spectrograph. This spectrograph used a novel transmission holographic grating and multiple element lenses to achieve high throughput, high resolution, and excellent imaging in a very compact package. The entire spectrometer measures about 5" wide, 17" long and 7" high. Within this package are both a holographic notch filter with associated lenses (which were external to the Chromex) and the spectrograph stage. The resolution attained by this spectrograph is superior to the Chromex and it is also smaller in size. The spectral coverage was 1550 cm^{-1} in a single exposure and is not adjustable without changing to a different grating (i.e. the grating is not rotatable). The fact that there are no moving parts can be beneficial for a robust system but limiting in flexibility.

Of the two spectrographs, the Kaiser has a much higher throughput due to its lower f-number. The measured gain in going from the Chromex to the Kaiser is a factor of 3.4 more signal with the Kaiser. The trade-off to the high resolution and high throughput is flexibility and the instrument response function. The transmission holographic grating in the Kaiser is not rotatable and therefore covers a fixed wavelength region. The Chromex grating is not only rotatable but there can be three gratings under computer control. Kaiser is developing a multiplexed grating that contains two transmission gratings in the same substrate. The gratings are angled so each directs its spectral information to different parts of the CCD (at the time of this work, we were not able to obtain this type of grating). In this way, a broader spectral region can obtained. One of the gratings covers the low wavenumber region and the other covers the high wavenumber region. The Kaiser spectrograph uses a fixed width slit. To change the slit width, the cover of the spectrograph has to be removed and a different width slit inserted. On the Chromex spectrometer, the slit is variable and can be adjusted under computer control. This makes varying the throughput and resolution by changing the slit width much more flexible.

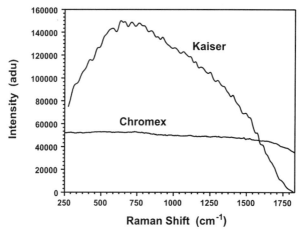

Figure 5   *Spectrograph system response for the Chromex and Kaiser spectrographs to a tungsten "white" light source*

In addition to the flexibility problem, transmission holographic gratings have a strong response function. This could be compared to a traditional grating with a strong blaze. Shown in Figure 5 is the spectrum of a tungsten bulb taken on both the Chromex and the Kaiser spectrographs. The Chromex was using the 600 g/mm grating blazed at 1000 nm. The spectral range in this figure is from 800 nm to 917 nm. It can be seen that the Chromex has an almost flat response over this region while the Kaiser exhibits a strong response function. This can be problematic especially when trying to use the water Raman band at 1640 cm^{-1} as an internal standard. The steep slope of the response distorts the spectrum. Some of this distortion can be corrected by using the "white light" source for flat fielding the spectrum. The problem is that in the long wavenumber region where the response is low (>1600 cm^{-1}), the noise component that is independent on the signal strength, is greatly amplified by the flat field correction, yielding poor signal to noise. The response peak of the Kaiser grating can be tailored to a specific region of interest.

2. 4 CCD Detector

The custom CCD detection camera used a back-thinned, AR coated, 1024 x 1024 pixel CCD (Tektronix, Model TK1024AB). The CCD has a peak quantum efficiency (QE) of 87% at 700 nm and the QE is >20% at 1000 nm with <5 electrons read noise. At the time this chip was acquired, it was the state of the art in terms of QE and read noise in the near infrared. Currently, this is now close to the standard specification for this model CCD (presently manufactured by SITe Inc.) and developments in the near future may lead to much better performance in the 800 nm -1000 nm wavelength region (see Future Developments). The CCD is mounted in a liquid nitrogen dewar (Infrared Laboratories, Model ND-1) and is typically operated at -105°C. For field use, the liquid nitrogen dewar can be replaced with a multistage thermoelectric cooler that is capable of maintaining -65 to -70 °C. This would cause a slight increase in the dark noise but the operation would not require the use and storage of liquid nitrogen. The camera is controlled by a

Photometrics (Model CE200A) camera electronics unit and a Photometrics (Model CC200) camera controller that is interfaced to a microcomputer (PC, 486-33). National Instruments LabVIEW® software is used for instrument control and data analysis. LabVIEW® is a graphical programming language and proved to be a very flexible and fast development tool for all aspects of the project. Software developed in our laboratory was used in all the instrument control and data analysis. This flexibility was essential for the evaluation of the many components and configurations of the instrumentation.

3 DETECTION RESULTS

3. 1 Benzene in Water and in Carbon Tetrachloride

Initial sensitivity tests performed with the 18-1 fiber-optic probe and the Chromex spectrograph have demonstrated unprecedented sensitivity for diode laser fiber-optic Raman systems. Benzene is often used as a standard in Raman spectroscopy due to its strong band near 992 cm^{-1} enabling comparisons of this work to sensitivities reported in the literature. For comparison to previous fiber-optic Raman studies in the literature where spectra of benzene were shown,[19] the same types of measurements were performed on solutions of benzene in carbon tetrachloride (CCl$_4$) and water. In this reference, a spectrum of benzene in CCl$_4$ at 0.05 M (2450 ppm) showed a RMS S/N≈17 (30 mW at the laser head and ~15 mW at probe at 783 nm, 75 μm slits, 300-s integration). Figure 6

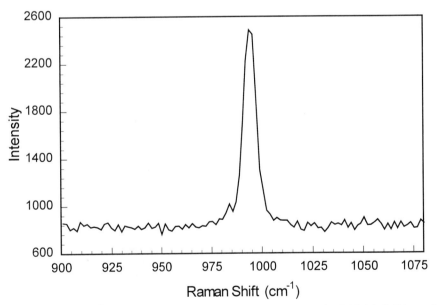

Figure 6 *992 cm^{-1} region of the fiber-optic Raman spectra of a 0.05 M (2450 ppm) sample of benzene in CCl$_4$. taken with a 10 second integration*

shows the same concentration (2450 ppm) taken on our instrument with a 10 second integration time. Here it can be seen that the S/N of the spectrum in this work is better (RMS S/N=60) even when the integration time is a factor of 30 less. In the work presented here, a RMS S/N of 17 is obtained at 0.0008 M (40 ppm) with 38 mW at 782 nm, 75 μm slits, and 300-s integration. This is an improvement of 62.5 times more sensitive (25 times with the laser power at the probe taken into account). Figure 7 shows the spectrum of 25 ppm benzene in CCl_4 as an example of the S/N that can be obtained in a 300 second integration at low concentrations. The limits of detection (RMS S/N=2) using the 992 cm^{-1} band of benzene were 1.2 x 10^{-4} M (6 ppm) in CCl_4 and 3.2 x 10^{-4} M (25 ppm) in water. The higher background in the Raman spectrum of water near the 992 cm^{-1} band led to a higher detection limit for benzene in water compared with benzene in CCl_4. Also reported in the literature was the detection of benzene in CCl_4 using a long-pathlength cell (1-m long) that yielded an enhancement of 40 over "conventional capillary sampling," producing a 2-ppm detection limit[28] using a more complicated sampling compared with the simple dip probe used in our work. The details of the experimental conditions under which this detection limit was obtained were not provided in the reference but it should be noted that the laser wavelength used was 488 nm giving a v^4 advantage of ~7.5 (992 cm^{-1} benzene band) over the present work using 782 nm excitation.

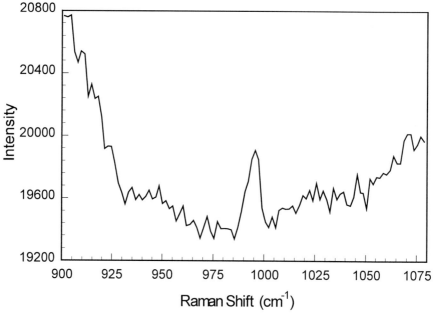

Figure 7 *Benzene Raman band from 0.0005 M (25 ppm) benzene in CCl$_4$ with an integration time of 300 seconds*

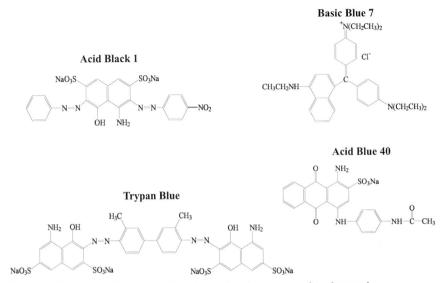

Figure 8 *Structural diagrams of the dye molecules examined in this study*

3. 2 Organic Dyes

The dye samples were obtained from the US. Environmental Protection Agency, Las Vegas, NV and had been purified by high performance liquid chromatography. A series of dilutions of a 1000-ppm stock solution was made for each dye using distilled, deionized water. Spectra were taken by dipping the 18-1 fiber-optic probe into the solution. The probe was rinsed with ethanol and dried between samplings. Exposure times for each spectral acquisition were limited to a maximum of five minutes. The structures of the dye molecules are shown in Figure 8.

The spectra of the four dyes are shown in Figure 9. The region shown, 850 cm^{-1} to 1650 cm^{-1} is the fingerprint region. These spectra are of 1000-ppm samples using a 60-s integration time. Acid black 1 and trypan blue have a higher Raman scattering cross section than acid blue 40 or basic blue 7, yielding far better detection limits. Note that the acid blue 40 and basic blue 7 have been scaled in Figure 9 by a factor of five. Absorption spectra of the dyes all show a strong absorption band near 600 nm, near zero absorbance at 700 nm, and no appreciable absorption at 782 nm (laser excitation wavelength). The excitation at 782 nm is far off resonance, and therefore resonant or preresonance enhancement of the Raman spectrum should not be an important factor in detection sensitivity. The spectra in Figure 9 show that although there are some similarities, each spectrum contains bands that could be used to differentiate the individual dyes. Also, it is important that each dye has several bands with sufficient intensity to allow identification even at low concentrations.

Table I shows a summary of the detection limits for the four dyes in this study. Two detection limits are given for each dye. The first limit is defined as that where the root mean square (RMS) signal-to-noise ratio (S/N) equals 2 on the strongest Raman

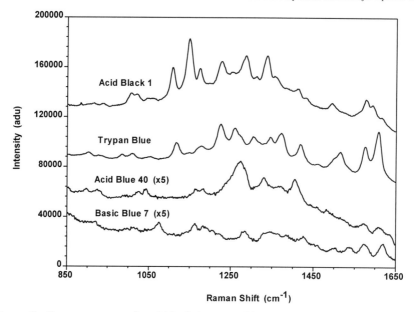

Figure 9   *Raman spectra of acid black 1, trypan blue, acid blue 40, and basic blue 7 at 1000 ppm in water. Note that the spectra of acid blue 40 and basic blue 7 are magnified by a factor of five*

band. The signal is measured by the peak height and the noise is the RMS noise at the baseline. Repeated measurements of low concentration solutions yielded reproducibility of peak height measurements of ±1% or better. The second detection limit is where at least six bands are identifiable (with S/N > 4) in the spectrum allowing for a better molecular identification. The integration time for these measurements was limited to 5 minutes. With longer acquisition times, lower limits could be achieved.

   For the measurements taken here, no internal standard was needed and the working curves were linear over the three orders of magnitude measured with correlations (r^2) of 0.999 for basic blue, 0.998 for acid blue, 0.996 for trypan blue, and 0.999 for acid

TABLE 1   *Diode laser/fiber-optic Raman detection limits. Concentrations are expressed in both ppm and molar units*

Compound	Detection Limit at RMS S/N=2 (strongest band)		Detection Limit (when 6 bands are observed)	
	ppm	Molar	ppm	Molar
Trypan Blue	0.4	4.16×10^{-7}	5	5.20×10^{-6}
Acid Black 1	0.2	3.24×10^{-7}	2	3.24×10^{-6}
Acid Blue 40	5	1.06×10^{-5}	50	1.06×10^{-4}
Basic Blue 7	25	4.86×10^{-5}	100	1.94×10^{-4}

black. In sampling where an internal standard is necessary, the 1640 cm^{-1} band of water provides a good internal standard. The laser power at the tip of the fiber-optic probe varied due to the condition of the diode laser and according to which laser was used: trypan blue (38 mW); acid black 1 (27 mW); acid blue 40 (26 mW); basic blue 7 (67 mW); the SDL Inc. diode laser was used only for basic blue 7. The initial diode laser used (Melles Griot), was operated for a period of time above the recommended output power. This shortened the lifetime of the diode and the power subsequently was lowered as noted above to extend its remaining life. Although the lifetimes of these diode lasers are quoted by the manufactures to be from 7,000 to 50,000 hours, they can be much shorter if the laser is run above the recommended output power.

Figure 10 shows spectra of acid black 1 (100-ppm and 20-ppm solutions) in its fingerprint region at various exposure times. Here it is demonstrated that spectra can be

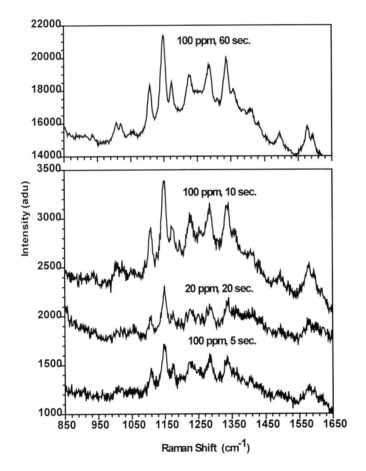

Figure 10 *Raman spectra of acid black 1 at 100 ppm (60-s, 10-s, and 5-s integration times) and at 20 ppm (20-s integration time)*

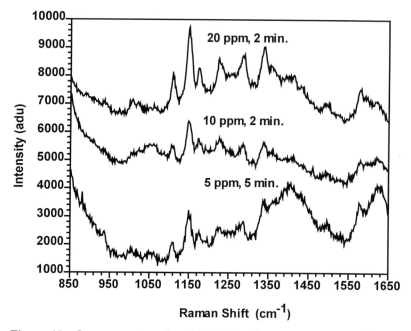

Figure 11 *Raman spectra of acid black 1 at low concentrations: 20 ppm, 2 min integration time; 10 ppm, 2 min integration time; 5 ppm, 5 min integration time*

obtained at low concentrations in very short exposure times with sufficient S/N to allow qualitative as well as quantitative determinations. This ability can be important in real-time monitoring applications. Figure 11 shows spectra of lower-concentration samples of acid black 1 taken with longer exposure times to demonstrate the S/N obtainable at these low concentrations. These spectra of acid black 1 were all taken with only 27 mW of laser power at the fiber-optic probe tip; all spectra shown in this work are raw (unsmoothed) data. With simple smoothing, as is often used in spectral analysis, the S/N could be greatly enhanced.

4 FUTURE DEVELOPMENTS

New diode lasers have recently become available that are stabilized and have powers of 300 mW at 785 nm (SDL, Inc.). This is a external cavity diode laser that was designed specifically for Raman spectroscopy due to an increased demand for such lasers. SDL, Inc. is also developing a lower power (100 mW range) stabilized laser that would be more cost effective and allow a less expensive Raman system to be built.

A new generation of charge injection device (CID) cameras will include a device developed especially for near-IR Raman spectroscopy. This device will be made with a deep depletion, thick epitaxial layer that will provide higher quantum efficiency in the 800 nm - 1000 nm region than is currently available in any commercial device. The expected increase is as much as a factor of two at 1000 nm (~40% QE). In addition, the new CID has the capability of random access collective reading. This allows any group of pixels on the device to be read collectively in a single read. Normally each pixel on a CID has to be read individually along with its associated read noise, which is much higher on a CID than a CCD. The often used technique to reduce the read noise on a CID is to take advantage of the ability to do non-destructive reads and average several re-reads of each pixel, thereby lowering the effective read noise. With the collective reading, which is also non-destructive, all the pixels in one column of the device (corresponding to a single wavelength in the focal plane) can be read together. In this particular device, the format is 1024 x 256 pixels. Therefore, the 256 pixels in a column can be read together with the same read noise associated with a single pixel read.

The ability to do non-destructive reads allows the spectrum to be monitored while the signal is being integrated, something not possible on a CCD. This is advantageous on long integrations since cosmic ray events that cause very high signals in a single pixel can be handled on a pixel by pixel basis as they happen. Also, the Raman signal can be monitored and the integration stopped when the signal is sufficiently strong. This is in contrast to a CCD, which can only be read at the end of a set integration time where numerous cosmic ray events can cause problems. This new CID camera will be tested in our laboratory in the fall of 1996.

5 CONCLUSIONS

This work has demonstrated that diode laser/fiber-optic Raman spectroscopy is capable of detection levels not usually associated with normal Raman scattering. These sensitivities have been achieved with normal Raman, without the problems and limitations involved in resonance Raman or surface-enhanced Raman. This, coupled with the ability of fiber-optics to provide remote as well as *in-situ* probing, make this Raman technique appropriate for many environmental and industrial monitoring applications. Even at an analyte concentration of 20 ppm, a spectrum with several identifiable peaks can be obtained in 20 s (see Figure 10). With the combination of the small relative size of the instrument, its sensitivity, and the spectral databases that are available, this should prove to be a versatile analytical field instrument. *In-situ* water analysis will eliminate hazardous waste created by conventional laboratory workup.

With further development of the fiber-optic probe technology and the possibility of more powerful diode lasers in the near future, this detection level can be improved. While advances in transmission holographic grating spectrographs have yielded very high throughput systems (*f*/1.8), they have limitations for general applications due to the limited wavelength coverage and the strong variation in the throughput across the wavelength region covered. Though current CCD detectors are quite sensitive in the near IR region used in this work, new CID devices that will be optimized for the near IR sensitivity and have novel readout capabilities should be beneficial for Raman spectroscopy.

References

1.   R. G. Luthy and M. J. Small, *Environ. Sci. Technol.* **24**, 1620 (1990).
2.   G. R. Trott and T. E. Furtak, *Rev. Sci. Instrum.* **51**, 1493 (1980)

3.   R. L. McCreery, M. Fleischmann, and P. Hendra, *Anal. Chem.* **55**, 146 (1983).

4.   S. Schwab and R. L. McCreery, *Anal. Chem.* **56**, 2199 (1984).

5.   K. Newby, W. M. Reichert, J. D. Andrade, and R. E. Benner, *Appl. Opt.* **23**, 1812 (1984).

6.   S. Schwab and R. L. McCreery, *Anal. Chem.* **58**, 2486 (1986).

7.   D. D. Archibald, L. T. Lin, and D. E. Honigs, *Appl. Spectrosc.* **42**, 1558 (1988).

8.   M. A. Leugers and R. D. McLachlan, in *Chemical, Biological, and Environmental Applications of Fibers,* R. A. Lieberman and M. T. Wlodarczyk (eds.) Proc. Soc. Photo-Opt. Instrum. Eng. **990**, 88 (1988).

9.   E. N. Lewis, V. F. Kalasinsky, and I. W. Levin, *Anal. Chem.* **60**, 2658 (1988).

10.   M. L. Myrick and S. M. Angel, *Appl. Spectrosc.* **44**, 565 (1990).

11.   M. L. Myrick, S. M. Angel, and R. Desiderio, *Appl. Opt.* **29**, 1333 (1990).

12.   K. P. J. Williams, *J. Raman Spectrosc.* **21**, 147 (1990).

13.   C. K. Chong, C. Shen, Y. Fong, J. Zhu, F. X. Yan, S. Brush, C. K. Mann, and T. J. Vickers, *Vibrat. Spectrosc.* **3**, 35 (1992).

14.   C.L. Schoen, T.F. Cooney, S.K. Sharma, and D.M. Carey, *Applied Optics*, **31**, 7707 (1992).

15.   C. Wang, T. J. Vickers, and C. K. Mann, *Appl. Spectrosc.* **47**, 928 (1993)

16.   J. M. Williamson, R. J. Bowling, and R. L. McCreery, *Appl. Spectrosc.* **43**, 372 (1989).

17.   S. M. Angel and M. L. Myrick, *Anal. Chem.* **61**, 1648 (1989).

18.   Y. Wang and R. L. McCreery, *Anal. Chem.* **61**, 2647 (1989).

19.   C.D. Allred and R.L. McCreery, *Appl. Spectrosc.* **44**, 1229 (1990).

20.   S.M. Angel, M.L. Myrick, and T.M. Vess, "Remote Raman Spectroscopy Using Diode Lasers and Fiber-optic Probes," in *Optical Methods for Ultrasensitive Detection and Analysis: Techniques and Applications*, (SPIE, Bellingham, Washington, 1991), Vol. 1435, pp. 72-80.

21.   S. M. Angel, T. M. Vess, and M. L. Myrick, "Simultaneous Multi-point Fiber-optic Raman Sampling for Chemical Process Control Using Diode Lasers and a CCD Detector," in *Chemical, Biological, and Environmental Fiber Sensors III*, (SPIE, Bellingham, Washington, 1991), Vol. 1587, pp. 219-231.

22.   C. D. Newman, G. G. Bret, and R. L. McCreery, *Appl. Spectrosc.* **46**, 262 (1992).

23.   D. A. Gilmore, D. Gurka, and M. B. Denton, *Appl. Spectrosc.* **49**, 508 (1995).

24.   M. M. Carrabba, K. M. Spencer, C. Rich, and D. Rauh, *Appl. Spectrosc.* **44**, 1558 (1990).

25.   M. J. Pelletier and R. C. Reeder, *Appl. Spectrosc.* **45**, 765 (1991).

26.   B. Yang, M. D. Morris, and H. Owen, *Appl. Spectrosc.* **45**, 1553 (1991).

27.   D. M. Pallister, K. L. Liu, A. Govil, M. D. Morris, H. Owen, and T. R. Harrison, *Appl. Spectrosc.* **46**, 1469 (1992).

28.   S. D. Schwab and R. L. McCreery, *Appl. Spectrosc.* **41**, 126 (1987).

APPLICATION OF A CCD TO PLANAR CHROMATOGRAPHIC ANALYSIS

Y. Liang, M. Bonner Denton

Department of Chemistry
University of Arizona
Tucson, AZ 85721

1 INTRODUCTION

Imaging high-performance thin-layer chromatography (HPTLC) has been shown to be a fast and efficient method for quantitative thin-layer chromatography (TLC).[1,2,3,4] Image acquisition allows the entire plate surface to be viewed simultaneously, which significantly decreases analysis time over slit scan detection and explores the high throughput of TLC to its maximum advantage.[5] Image acquisition can be useful in attempts to automate the analysis of TLC plates in that it can minimize human intervention and facilitate the extraction of quantitative information. Despite their inherent advantage, early versions of photo-diode and vacuum tube multi-channel detectors lacked the sensitivity and dynamic range of discrete photomultipliers.[6] More recent developments in solid state area array imagers have significantly improved multi-channel detection. Scientifically operated (cooled, slow-scan) charge coupled device (CCD) array detectors have demonstrated extremely low dark current and read noise characteristics while simultaneously providing high sensitivity from the soft x-ray to the near IR and wide dynamic range. These features have made the scientific CCD a nearly ideal detector for many low photon flux imaging applications, such as fluorescence HPTLC detection.[6]

The application of the CCD array detector in quantitative HPTLC analysis was explored by using aflatoxins as model compounds (Figure 1). Aflatoxins were selected because of their extremely toxic and carcinogenic effects. Strong correlation has been found between the levels of aflatoxins in human diets and incidence of liver cancer or death.[7] In addition, these compounds have been found on a variety of agricultural products, livestock feeds and commercial foodstuffs. While improvements in storage conditions have done much to reduce the problems associated with aflatoxin contamination of agricultural products, extensive routine monitoring for these compounds at all levels of production from growing, harvesting, storage and processing are required to control this severe health hazard.[8] Thus, a low cost and reliable technique for the fast and efficient analysis of large numbers of potentially contaminated samples is necessary for the practical routine monitoring of these hazardous compounds. A requirement of this technique should also include a high degree of sensitivity for the aflatoxins under investigation in order to determine contamination at very low levels.

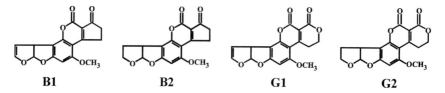

B1   **B2**   **G1**   **G2**

Figure 1 *Molecular structures of aflatoxin B1, B2, G1, and G2*

2 EXPERIMENTAL

Aflatoxin standards were obtained from Sigma Chemical Co. (St. Louis, MO) in crystal form. A laser grade fluorescence dye, rhodamine 6G (R6G, Kodak, Inc.), was utilized for system evaluation. Ground peanut butter and creamy peanut butter were bought from grocery stores. Standard solutions were prepared by dissolving the crystal standards in chromatographic grade methanol. The Association Official of Analytical Chemists (AOAC) recommended BF method[9] was used for the aflatoxins extraction from peanut butter samples. Three mobile phases were used for chromatographic development: anhydrous ethyl ether, chloroform / ethyl ether (7:3 v/v) or chloroform / acetone (9:1 v/v). Before sample application, all HPTLC plates were washed chromatographically with methanol and dried in air. All images were bias and flat field corrected.

The imaging system utilized in this study is shown in Figure 2. The CCD camera system was provided by Photometrics, Ltd. (Tucson, AZ) and consisted of a CC200 camera controller, a CE200 camera electronics unit and a PM512 CCD camera. The CCD was cooled to approximately -100°C with the use of liquid nitrogen in order to reduce the

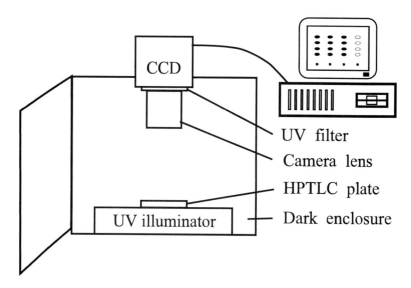

Figure 2 *Schematic diagram of HPTLC imaging system*

production of thermally generated charge to negligible levels. Because of the fluorescent nature of aflatoxins, fluorescence detection was used. A 400 nm high pass filter (CVI Laser Corp., Albuqueruqe, NM) was mounted on the front of the camera lens to achieve the proper emission filtering characteristics for the aflatoxins under investigation. Ultraviolet illumination of the HPTLC plates was accomplished with a transilluminator which contained two 8 watts 365 nm UV light tubes. The transilluminator was overlaid with a UV-340 nm glass filter. The whole system was controlled by microcomputer using National Instruments LabVIEW software for instrument control and data analysis.

3   RESULTS AND DISCUSSION

3.1 System Evaluations.

The development of new solid state imaging systems for HPTLC has placed considerable demands on the quality and reproducibility of thin-layer plates, and on the accuracy and precision of the sample applicator and the imaging system.

3.1.1 Quality of HPTLC Plates. It has been observed in our previous work that commercially manufactured plates have a high degree of fluorescent impurities in or on the support material that significantly affect the quality of the measurements[6]. The sporadic occurrence of contamination spots represented the ultimate limiting factor in the sensitivity of the quantitative HPTLC analysis, not the noise associated with the detector or the excitation source.[10]   The nature of the fluorescence background on different brand HPTLC silica gel plates have been investigated.

Experimental results showed that fluorescence from the glass substrate accounts for 34.6% to 80.3% of the total background fluorescence and over 30 % of the background noise level (Table 1), except the MN plate in which the major noise source was from individual impurity spots on the plate. This problem, intrinsic to the glass, could be reduced by substituting the glass backing with low or non fluorescent materials, such as quartz. The silica layer contributes two different types of fluorescence background: an overall background and the individual fluorescent spots. The individual fluorescent spots were observed to be the greatest contributor to the variation in the fluorescent background. These spots are caused by fluorescent impurities adsorbed in or on the silica sorbent layer. The existence of some fluorescence impurities on sorbent layers is expected due to the many possible sources of contamination during plate manufacturing, packaging and handling processes. Part of the fluorescent impurities can be removed by cleaning the plate with appropriate solvent systems. If the plates are not cleaned prior to use, these impurities can be redistributed to a higher position on the plate causing a characteristic rising baseline in the direction of chromatography. The fluorescent background and noise, which could not be removed by plate cleaning procedures, can be adjusted mathematically with a flat field correction method.[6] The improvement of the flat-field correction toward the reproducibility and the sensitivity of the HPTLC quantitative results can be seen clearly from Figure 3. Therefore, plate pre-cleaning to minimize the adverse effects of impurities in the sorbent layer and flat-field correction to normalize the spatial distribution of the plate background and the excitation source are required to obtain reliable quantitative results in HPTLC analysis. Comparing with the

Table1. *Fluorescence background and noise level of different precoated HPTLC plates.*

Plates	HPTLC Plate				Glass		$I_{glass}/I_{plat}*100$ (%)		Impurities
	Before cleaning		After cleaning						
	Avg. Int.[a]	SD[b]	Avg. Int	SD	Ave. Int.	SD	Before	After	removed
MN	804	84.4	589	29.6	204	4.59	25.4	34.6	yellow
AN	1514	51.0	1488	34.7	770	13.6	50.9	51.7	invisible
EM	437	14.3	288	6.34	227	4.72	51.9	78.8	pale yellow
WM	817	18.6	687	13.9	552	10.7	67.6	80.3	yellow

(a) Ave. Int. : average fluorescence intensity in arbitrary units.
(b) SD :   standard deviation of the fluorescence intensities based on the whole image.

effects of the fluorescent background, the contribution of the scattering loss to the sensitivity of the detection is negligible.

3.1.2 Sample Applicator. Most sample application is accomplished by the use of high precision pipettes which can measure samples in nanoliter volumes. This method of application is highly technique oriented, and can result in poor precision of quantitative transfer and less than optimal quality of the chromatographic separation. An aerosol spray-type sample applicator was constructed and used for quantitative sample delivery (Figure 4).[11] Nitrogen gas was used to disperse the solution into fine drops and to facilitate solvent evaporation. The system applied 0.5μl sample volumes to the silica gel plate in a very focused and controlled manner. The sample spot size on the plate was smaller than 1 mm in diameter.

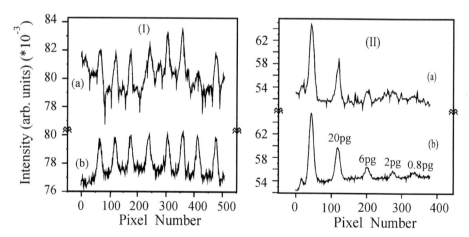

Figure 3   *Effects of the flat field correction to the quantitative results. (I) reproducibility test of 10 pg R6G, (II) LOD test of R6G, before (a) and after (b) flat field correction.*

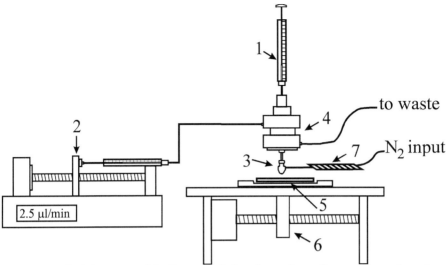

Figure 4 *Block Diagram of the "micro nebulizer" sample applicator. 1)sample syringe, 2) infusion pump, 3) modified concentric flow nebulizer, 4) 0.5 μl HPLC sample injection loop, 5) TLC plate and holder, 6) plate movement stage, 7) heating tape.*

3.1.3 System Performance. The system performance was evaluated by using R6G standard solutions. Seven to eight spots of the same quantity of R6G standards were applied onto each 5 x 5 cm HPTLC plate. The integrated fluorescence observed with the scientific CCD camera of the single component R6G samples was measured before and after chromatography. The results of the inter-spot precision are shown in Table 2. The pooled relative standard deviation before and after plate development was less than 1.4%. The accuracy of the system was investigated by spotting six different R6G standards, in the range of 0.8 pg to 80 pg, onto each 5x5 cm HPTLC plate with the 20 pg spot as a test sample. The recovered quantity of the test standard was obtained from the calibration curve derived from the other five R6G standards. The results showed that the mean recovery percentage of the test standard was 99% with relative standard deviation of 2.1% (average of six trials).

Table 2 *Inter-Spot Precision Results*

Mass (pg)	# of Samples	%RSD before chromatography	%RSD after chromatography
6.00	7	2.04	2.08
20.00	7	1.48	1.52
50.00	8	.59	0.78
100.0	7	.77	1.32
200.0	7	1.25	1.53
Pooled %RSD		1.35	1.40

* Data obtained by measuring the total integrated fluorescence from each spot.

Table 3 *Detection Limits of Aflatoxins*

Aflatoxins	HPTLC-CCD (pg)	HPLC-MS[12] (pg)
G2	3.3	10
G1	4.9	5
B2	3.0	4
B1	4.5	4

The results of the inter-spot precision and the recovery percentage of the spotted R6G standard indicate that a reasonable level of precision and accuracy is obtainable for quantitative HPTLC analysis using scientific CCD detection.

3.2 Analysis of Aflatoxins

3.2.1 Dynamic Range and Sensitivity. A series of standard aflatoxin mixtures were applied to a 5x10 cm HPTLC plate. Chromatographic spectra were obtained from the images for each of the aflatoxin species and for each of the samples. The average limits of detection (LOD) based on a signal to noise ratio (S/N) of 3 for the four aflatoxins were found to be between 3 to 5 pg (Table 3). The LOD results obtained by HPTLC with CCD detection were close to or lower than those reported by Kussak *et al*[12] for aflatoxins using HPLC Mass Spectrometry. Good linear fits to the data ($p<0.001$, $r^2>0.96$) were achieved over greater than two orders of magnitude for all the aflatoxins under investigation.

3.2.2 Peanut Butter Study. Aflatoxins are extracted and condensed from peanut butter samples. The repeatability of the extraction method and the imaging system was assessed by measuring the percent recovery of samples spiked with aflatoxins. Twenty-five gram samples of aflatoxin free peanut butter were spiked with 250 ng of B1 & G1 and 75 ng of B2 & G2, respectively. Figure 5 shows one of the HPTLC images containing the spiked peanut butter extracts. No detectable levels of aflatoxins were found in the unspiked sample (as indicated in Figure 6). The aflatoxin free samples were freshly ground peanuts without any additives. The images show that except for the aflatoxin species most of the extracted components were left at the origin of sample application, suggesting that a sample precleaning step is required if HPLC is to be used for analysis. The recoveries of the four aflatoxins are in the range of 90 to 100% with acceptable standard deviations (less than 2.8%, Table 4).

Table 4 *Recovery of Aflatoxins from Spiked Peanut Butter Extracts*

Aflatoxins	Added (ng/g)	Recovery* (%)	RSD *(%)
B1	9.84	97.85	2.63
B2	2.95	90.38	2.78
G1	9.84	100.2	1.71
G2	2.95	98.75	2.47

* Recovery percentage and RSD are average values of six assays.

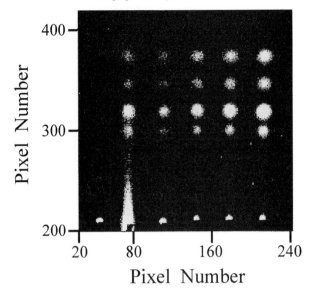

Figure 5   *CCD image of the separated aflatoxins with spiked peanut butter extracts. From top to bottom:B1, B2, G1, G2 and sample origin. From left to right: pure solvent, peanut butter extracts, standards with increased concentrations.*

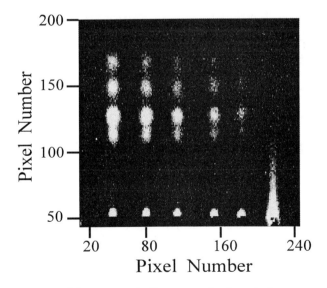

Figure 6   *CCD image of the separated aflatoxins with aflatoxin free peanut butter extracts. From right to left: peanut butter extracts, standards with increased concentration.*

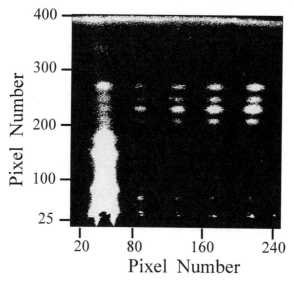

Figure 7   *CCD image of the separation of the aflatoxins with contaminated peanut*
butter extracts. Same sequence as in Figure 6.

A commercial creamy peanut butter product was also analyzed. Approximately 3 ppb level of total aflatoxins were found in the peanut butter extracts (Figure 7). The aflatoxin content was determined to be much lower than the FDA regulation level, which is recommended at 20 ppb. The existence of the co-developed interfering compounds caused a characteristic decreasing baseline in the direction of chromatography, resulting in poor precision on the quantification of the G2 component.

4 CONCLUSIONS

The scientific CCD based imaging system described here provides excellent sensitivity, accuracy, precision and linearity for the quantitative determination of aflatoxins on HPTLC plates. The detection limits of the method are limited by the fluorescence background and noise of the TLC plates, but not noise associated with the detector or the excitation source. The ability of this technique to separate and detect samples in parallel can result in significant reduction in analysis time over techniques utilizing sequential separation or detection. The detection limits for each of the aflatoxins were less than 5 pg, making this system ideal for easily identifying contaminated products. As low as 3 ppb of aflatoxins were detected from naturally contaminated peanut butter. The flexibility of the system makes it readily adaptable to many screening or routine monitoring applications requiring high sample throughput and sensitivity.

References

1   J. A. Cosgrove and R. B. Bilhorn, *J. of Planer Chromatogr.* 1989, **2**, 362.
2   S. M. Brown and K. L. Busch, *J. of Planer Chromatogr.* 1992, **5**, 338.
3   F. G. Sánchez, A. V. Díaz and M. R. F. Correa, *J. Chromatogr.* 1993, **A655**, 31.
4   E. H. Jansen, D. Van Den Bosch and R. W. Stephany, *J. Chromatogr.* 1989, **489**, 205.
5   J. C. Touchstone, 'Practice of Thin Layer Chromatography', 3rd edition, A Wiley-Interscience publication, John Wiley & Sons, Inc., 1992.
6   M. E. Baker and M. B. Denton in 'Charge Transfer Devices in Spectroscopy', J. V. Sweedler; K. L. Ratzlaff and M. B. Denton (eds.), VCH publications, Inc. 1994, pp 197.
7   I. Dvorackova, 'Aflatoxins and Human Health' CRC Press, Inc., 1990.
8   B. T. Hunter, *Consumers' Research* 1989, *June*, pp8.
9   P. M. Scott, 'AOAC Official Methods of Analysis', AOAC Arlington VA, 1990, Chap. 49 Natural Poisons, (970.45)
10   Y. Liang; M. E. Baker; D. A. Gilmore and M. B. Denton *J. of Planar Chromatogr.* 1996, **9**(4) in press.
11   Y. Liang; M. E. Baker; B. T. Yeager and M. B. Denton, submitted to *Analytical Chemistry*.
12   A. Kussak; C. A. Nilsson; B. Andersson; J. Langridge, *Rapid Communications in Mass Spectrometry* 1995, **9** (13), 1234.

FIBER-OPTIC ARRAY SENSORS

David R. Walt,* Paul Pantano, Karri L. Michael, and Anna Panova

The Max Tishler Laboratory for Organic Chemistry
Department of Chemistry
Tufts University, Medford, MA 02155, USA

1 INTRODUCTION

Despite many innovations and developments in the field of fiber-optic chemical sensors, optical fibers have not been employed to both view a sample and concurrently detect an analyte of interest. While chemical sensors employing a single optical fiber or non-coherent fiber-optic bundles have been applied to a wide variety of analytical determinations, they cannot be used for imaging. Similarly, coherent imaging fibers have been employed only for their originally intended purpose, image transmission.

We describe here a new technique for viewing a sample and measuring surface chemical concentrations that employs a coherent imaging fiber. In addition, we present preliminary in situ results whereby these imaging fiber sensors are being used to monitor the onset of corrosion and neurochemical dynamics. Finally, we discuss progress toward the fabrication of a near-field optical array.

1.1 Combined Imaging and Chemical Sensing

An optical imaging fiber is comprised of thousands of individual 3-4-μm diameter optical fibers melted and drawn together in a coherent manner such that an image can be carried and maintained from one end to the other (Figure 1).[1] In the present innovation, a 350-μm-diameter distal fiber surface, which contains approximately 6000 optical sensors, is coated with a uniform, planar sensing layer that can measure chemical concentrations with spatial accuracy, yet is thin enough so that it does not compromise the fiber's imaging capabilities. By combining the distinct optical pathways of the imaging fiber with the spatial discrimination of a charge coupled device, visual and fluorescence measurements can be obtained with 4-μm spatial resolution over tens of thousands of square microns. Biological samples imaged with these planar, polymer-modified imaging fibers have included sea urchin eggs (50-100 μm diameter) and mouse fibroblast cells (10-20 μm diameter).[2] The intended use of these modified imaging fibers is significantly different from that of microscopy; the ultimate goal is to use these sensors to make measurements in remote locations. Such optical sensor arrays reduce the precision required to position

Figure 1   *Image of a 350-μm-diameter imaging fiber (comprising ~6000 optical fibers*
each with a diameter of 3-4 μm) as acquired by a CCD camera. Far-field
viewing is made possible by a distal GRIN lens.

an extremely small probe and offer major advantages over microelectrode arrays in both
ease of fabrication and manipulation at the cellular level.

   1.1.1 Imaging the Initial Phases of Corrosion. According to the local cell theory of
corrosion, the anodic dissolution of a metal at certain locations on a surface occurs with a
corresponding cathodic reaction at another surface location(e.g., the reduction of oxygen
which generates hydroxide).[3] A pH sensor was employed to image the hydroxide
generated at the surface of a copper electrode. The pH sensor was fabricated by spin-
coating a N-fluoresceinylacrylamide-derivatized hydroxyethyl methacrylate polymer onto
the distal face of an imaging fiber.[2] The fluorescence intensity of the immobilized pH
indicator is enhanced upon deprotonation. The pH sensor was placed in contact with a
copper electrode and fluorescence images were acquired before and after the application
of a corrosion-initiating voltage (Figure 2, top). The mean fluorescence intensity from a
region on the copper surface was chosen and plotted versus time (Figure 2, bottom). The
fluorescence intensity rises dramatically as the potential (-1.0 V vs. Ag/AgCl) is applied
(suggesting the production of hydroxide). Furthermore, some regions of the copper
surface are obviously more reactive than others.

   1.1.2 Imaging Neurochemical Dynamics. The specificity of optical pH sensors can be
extended dramatically by coupling indicator dyes with enzymes.   An acetylcholine-
sensitive layer was created by first co-immobilizing acetylcholinesterase (AChE) in a
water-soluble,   functionalized   prepolymer   known   as   poly(acrylamide-*co*-N-
acryloxysuccinimide).[4] The enzyme-derivatized polymer was spin-coated onto an imaging
fiber and subsequently reacted with fluorescein isothiocyanate. In this arrangement, the
dissociated protons from the enzyme-generated acetic acid quench the fluorescence of the

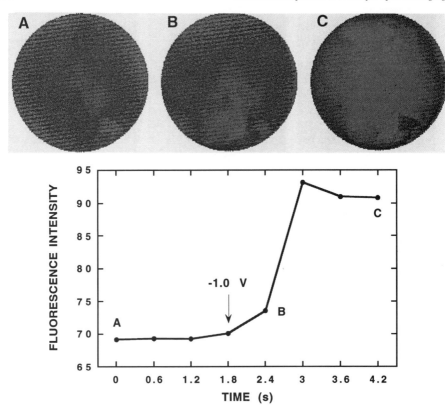

Figure 2   *Fluorescence image of a copper electrode placed in a 5 mM phosphate buffer solution as observed through a pH-imaging fiber sensor before (A) and after (B-C) the application of a corrosion-initiating voltage. The integration time of the charge coupled device was 300 ms; high fluorescence intensities are denoted by lighter shading.*

immobilized dye in proportion to the concentration of acetylcholine (ACh). The acetylcholine biosensor has a detection limit of 35 μM and a fast (<1 s) response time.[2]

In association with Professor Barry A. Trimmer of the Department of Biology at Tufts University, we are using these AChE-modified imaging fiber sensors to measure the stimulated release of ACh from the planta-hair afferents of the tobacco hornworm, *Manduca sexta*. Figure 3A shows a white light image of a caterpillar ganglion as observed through an AChE-modified imaging fiber sensor. The fluorescence images of the same region as viewed through the AChE-modified imaging fiber sensor are shown before (Figure 3B) and after (Figure 3C) electrical stimulation. The top right quadrant of the image shows a significant quenching of fluorescence where the sensory neurons are believed to terminate. The mean fluorescence intensity from this region was plotted versus time (Figure 3, bottom) and shows the fluorescence intensity decreases only when the neurons are electrically stimulated, and the fluorescence begins to recover as soon as

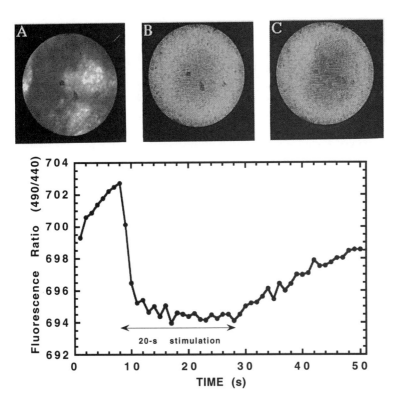

Figure 3 *(A) White light image of a caterpillar ganglion as observed through an AChE-modified imaging fiber sensor. The fluorescence image of the same region from the through an AChE-modified imaging fiber sensor is shown before (B) and after (C) electrical stimulation. (Bottom) The mean fluorescence intensity from top right quadrant is plotted versus time. The integration time of the charge coupled device was 300 ms; high fluorescence intensities are denoted by lighter shading.*

the stimulation was halted. Obviously, much more work is needed to prove unequivocally that the released chemical is acetylcholine, but, these data demonstrate the ability to correlate dynamic chemical events to cellular morphology.

1.2 Towards the Fabrication of a Near-Field Optical Array

Near-field Scanning Optical Microscopy (NSOM) is based on the fact that although light cannot be focused to a spot less than one-half the wavelength of light ($\lambda / 2$), it can be directed through an aperture with dimensions smaller than $\lambda / 2$. When light is directed through a sub-λ-sized hole, the portion that passes through the hole will at first be confined to the dimensions of the aperture before it rapidly diffracts in all directions. If a sample is brought within the near-field of a sub- $\lambda/2$ -sized aperture, and the light is raster-scanned over the sample, a two-dimensional image can be created in a serial fashion (one point at a time). [5-8]

Obviously, the easiest way to produce a sub- $\lambda/2$ -sized aperture is to drill a hole in a piece of metal. The problem with this approach is that one will only be able to image samples that are atomically-smooth. The first NSOM apertures suitable for real-world-samples were metal-coated glass pipettes. In this approach, one simply takes a small glass capillary tube and pulls it while heating. In this way, the internal diameter of the glass pipette can be pulled down to 50 nm. A thin film of an opaque metal (such as aluminum) is then evaporated along the outside walls of the pipette in such a way as to leave a small transparent window or aperture only at the tip. Light is directed through this aperture by sending the output of a laser beam through a small optical fiber placed in the non-pulled end of the pipette. [7] Today, single-mode optical fibers can be pulled in order to reduce the loss of light at the glass capillary tube / optical fiber junction. After the cladding material of the optical fiber is manually removed with a razor blade, the fiber is heated by a carbon dioxide laser and pulled. The side walls of the fiber tip are then coated with aluminum as described above such that an aperture as small as 10-20 nm is produced and resolution on the order of 15-50 nm can be achieved.

In order for an imaging fiber to be used to create an array of near-field apertures, several issues must be addressed. First, the clad of each individual optical fiber must be

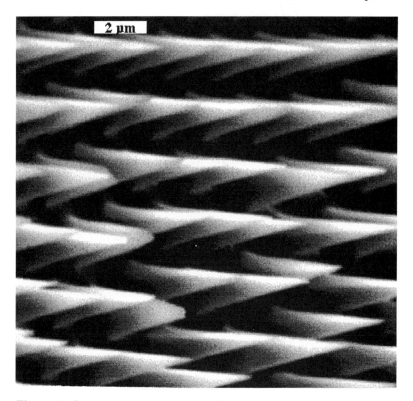

Figure 4   *Scanning electron micrograph of a chemically-etched imaging fiber*

removed, and second, the diameter of each individual optical fiber must be decreased to a dimension less than $\lambda/2$. Research in our laboratory has led to a development that solves both of these issues. Figure 4 is an electron micrograph showing an imaging fiber that has been chemically etched. The cladding material surrounding each individual optical fiber has been removed, and the tips of each optical fiber are now on the order of 200 nm. It should be noted that with this array architecture, one would not have to raster scan to build up an image (as is done with a single fiber-optic aperture); rather, only the region in between neighboring optical fibers would need to be scanned.

Recently, we have coated these etched imaging fibers with an opaque coating of aluminum. Work is in progress to remove the aluminum only at the tip of each tapered optical fiber so as to create an array of sub-$\lambda/2$-sized apertures. The success of this stage of the project will be evaluated with a scanning tunneling microscope.

ACKNOWLEDGMENTS

We gratefully acknowledge the National Institutes of Health (GM-48142) for financial support of this work.

References

1. P. Pantano and D. R. Walt, *Anal. Chem.* 1995, **67**, 481A.
2. K. S. Bronk, K. L. Michael, P. Pantano and D. R. Walt, *Anal. Chem.* 1995, **67**, 2750.
3. R. C. Engstrom, S. Ghaffari and H. Qu, *Anal. Chem.* 1992, **64**, 2525.
4. A. Pollak, H. Blumenfeld, M. Wax, R. L. Baughn, G.M. Whitesides, *J. Am. Chem. Soc.* 1980, **102**, 6324.
5. A. Lewis in "New Techniques of Optical Microscopy and Microspectroscopy", R. J. Cherry, Ed.; MacMillan Press: 1991; Chapter 2.
6. D. W. Pohl in "Advances in Optical and Electron Microscopy", T. Mulvey, and C. J. R.Sheppard, Eds.; Academic Press: 1991;243-312.
7. A. Lewis and K. Lieberman, *Anal. Chem.* 1991, **63**, 625A.
8. T. D. Harris, R. D. Grober, J. K. Trautman, E. Betzig, *Appl. Spectrosc.* 1994, **48**, 14A.

DEPTH RESOLVED ABSORPTION MEASUREMENTS IN SCATTERING MEDIA BY EXPERIMENT AND MONTE CARLO SIMULATION

William F. Long, Lorenzo Leonardi, and David H. Burns

Department of Chemistry
McGill University
Otto Maass Chemistry Building
801 Sherbrooke St. West
Montréal, Québec H3A 2K6

ABSTRACT

The feasibility of depth resolved absorbance determinations in human tissues based on diffuse reflectance measurements is investigated. Lateral measurements of intensity were made to obtain multi-perspective information through a specimen. Both Monte Carlo simulations and experimental measurements of known phantoms were used to evaluate the sensitivity of the method to changing sample absorption. Experimental diffuse reflectance distributions were obtained from a collimated incident source and a layered absorbing/scattering sample. A fiber optic detection system recorded lateral responses as a function of phantom composition. From the model, the "most probable path" distributions of photons reaching a detector located on the surface were determined. These distributions suggest that as the detection point is moved further away from the source, information regarding deeper layers may be obtained. This depth resolved tomographic technique may, in the future, provide a means in which to study spatial relationships of chromogenic constituents in scattering media.

1 INTRODUCTION

The non-invasive investigation of the interaction of light with tissue has many applications in diagnostic medicine. Bio-energetic status measurements can be made, for example, in the 700-1300 nm wavelength region. Here, hemoglobin, myoglobin, and cytochrome aa_3 are strongly absorbing and have oxygen-dependent absorption coefficients.[1-3] The total attenuated light can be measured as the steady state transmittance or remittance. Relative changes in transmitted light intensity at specific wavelengths can, therefore, be used to measure tissue oxygenation. In clinical settings, transmittance measurements are typically used where the path length through the sample is short, such as a finger, earlobe, or toe, since the light must penetrate through the sample. However, transmission-based oximetric techniques measure average capillary oxygenation in a single region in space and are limited to thin portions of the body. With specialized equipment it has been shown to be possible to use spectroscopic measurements in larger tissue samples. For example, Cope and Delpy have measured light transmission through the neonatal skull to continuously monitor cerebral oxygenation in infants.[4]

In a clinical setting, the reflectance or remittance geometry is preferred for bio-spectroscopic measurements. This method also provides a non-invasive technique for the analysis of muscle, heart, and brain tissue. Remittance measurements involve irradiating the sample and observing reflected light. In this case, a small fraction of the light is scattered in such a manner that it enters the detector located some lateral distance away from the source. The depth of the penetrated light is unclear, as it is dependent on the source/detector distance.[4,5] In general, the further the detector is located from the source, the larger the depth of penetration and the lower the intensity measured.

In this study, experimental measurements and computer simulations of light penetration through an absorbing/scattering media were performed in the remittance geometry. The results demonstrate how this information can be used to study depth resolved absorption estimates in thick samples.

2 BACKGROUND

Optical properties of tissue can be described by two basic processes, absorption and scattering, which account for the total attenuation of light through a sample. A schematic representation of the process is shown in Figure 1. When light enters a sample, a portion is absorbed by the analyte dispersed through the cellular and intercellular space. For this portion of the attenuation, the Beer-Lambert relation should hold and the exponential attenuation of light through a sample would be observed. For measurements made in the near-infrared spectral region, the absorption by tissue is small,[6,7] with typical values for the absorption coefficient, μ_a, between 10-100 mm^{-1}. The major absorptions in the near infrared are from low lying electronic excitation in chromophores such as hemoglobin, myoglobin and cytochrome and from overtones of molecular vibration and combination bands due to OH, CH and NH stretching and bending modes.

Because of the relatively small absorption cross-sections in tissue in the near infrared, scattering becomes a major contribution to the attenuation of the light. Scattering can occur due to refractive index variations in the different components in tissue or by elastic scattering such as in the case of light interaction with collagen in the cell membranes. Typical scattering coefficients, μ_s found in this wavelength region for tissue lie in the range

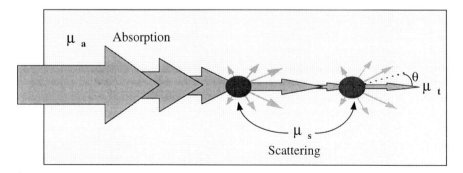

Figure 1 *Cross-sectional view of the scattering and absorption interaction with light. As light penetrates through the sample the intensity is reduced due to the absorption. Light is also attenuated by species located in the tissue which scatter the light.*

of 0.01-1 mm^{-1}. Likewise, the magnitude of the scattering cross-section in the tissue is observed to be inversely proportional to the wavelength of light. For soft tissue such as capillary beds and skin in the finger, a $\lambda^{-2.55}$ relationship of the μ_s is found.[1]

A single photon entering a tissue sample will experience many scattering events as it travels through the tissue. The distribution of the scattered photons can be described by a phase distribution function $P(\theta)$, where θ is the angle of the scattered photon after the interaction with a single scattering event. The mean cosine angle of the phase distribution is used to describe the anisotropy of the scattered photon and is defined as g. The values for g range from -1 to 1. A value of -1 represents a totally back scattered event in which the photon is redirected back to the source, and a value of 1 represents a totally forward scattered event in which the photon does not deviate from its original path. Likewise, at a value of 0 the distribution is isotropic. Typical values of g for tissue in the near infrared wavelength region are from 0.8 to 0.99[1]. The large g values imply that even in relatively highly scattering samples, the light is to a great degree forward directed. Scattering anisotropy can be combined with the scattering coefficient to produce a single term, namely the reduced scattering coefficient: $\mu_s' = \mu_s (1-g)$. Therefore the total measured attenuated light is the sum of the absorption and scattering contributions: $\mu_t = \mu_a + \mu_s'$.

3 COMPUTER SIMULATIONS

To estimate the distribution of photons in a scattering media such as tissue, Monte Carlo simulations of the steady state photon distribution in a slab of varying absorbance, μ_a, and a constant scattering coefficient, μ_s, were performed. The slab comprised four parallel layers with respect to the surface, each with a randomly chosen absorption coefficient (0.17, 0.35 or 0.70 cm^{-1}). The values of the scattering coefficient and the scattering phase functions were chosen to be similar to those found in human tissue.[1] The phase function, approximated by the Henyley-Greenstein function, was varied by the parameter, g, (0.95 - 0.99). The form of the Henyley-Greenstein function by Keijer was used to give a probability function for use with numerical methods.[8] A photon's free path, l, between each scattering event was chosen randomly based on the expression $l = \exp(-\mu_s)$.[9] Attenuation due to absorption was treated classically according to the Beer-Lambert relationship: $I = I_0 \exp(-\mu_a l)$.

The domain of the problem was fixed to a 1.5 cm lateral distance in both directions away from the source and to a depth of 1 cm. Simulations were started with all photons traversing perpendicular to the surface with the light injection equidistant from the boundaries. Boundary effects were not accounted for in this simulation (*i.e.* photons were followed until they either escaped or were detected). The simulation was allowed to proceed for a total time duration of 100 picoseconds. This ensured that all photons could traverse the 1 cm slab with multiple scattering. The steady state distribution was approximated by integrating over all calculated time dependent distributions of a 50,000 photon pulse.

Information from each simulation included the steady state distribution and the back-scattered light intensity at the surface as a function of lateral distance from the source. Because of symmetry, the two distributions on each side of source were averaged and the result was smoothed to reduce the noise inherent in the statistical nature of the modeling technique.

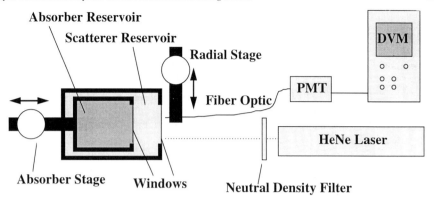

Figure 2 *Back-scattering experimental setup*

4 EXPERIMENTAL WORK

The experimental apparatus used to obtain the lateral distribution is shown in Figure 2. A 5.8 mW HeNe laser attenuated by a 1.0 O.D. neutral density filter served as the source. The detection system used was a 1.0 mm diameter fiber optic attached to a photomultiplier tube detector. A digital volt meter recorded the output signal. A 1:1 water to 10% milk fat cream solution was chosen as the scatterer and was present in both black reservoirs. For the absorber, 5 drops per 25 ml of a blue-green water-based dye was added to the movable reservoir to obtain a greater than 2 OD. The windows into the reservoirs were made of 0.34 mm glass. Detector responses were recorded as a function of the lateral position of the fiber optic (2-18 mm) and as a function of absorber position (0.25, 0.50 and 0.75 cm). An experiment with no absorber present was also done.

5 RESULTS AND DISCUSSION

For remittance measurements, one point on the surface is illuminated and the diffusely reflected light is collected some distance away. The light collected will have gone some average path through the sample. An idealized schematic for the light path is shown in Figure 3. For the four-layer sample in Figure 3, the light path will travel through each of the layers with some average distance and will have a weighted pathlength for the overall attenuation. To determine the weightings for the path lengths in each of the layers, it is important to determine the probability distribution of the light at each of the depths for a given source detector position. From this information the effect of absorption in a given layer may be determined from the measured attenuation at the surface. The results of a Monte Carlo simulation can be used to estimate the probability distribution of the measured light for a given source and detector position. In particular, the Monte Carlo simulation described in the methods section provides an estimate of the steady state distribution of light in a sample for a given source position. When the intensity of the source is taken to be unity then each point in the distribution can be visualized as the fraction of all the initial intensity reaching a certain depth in the sample. Likewise, the light

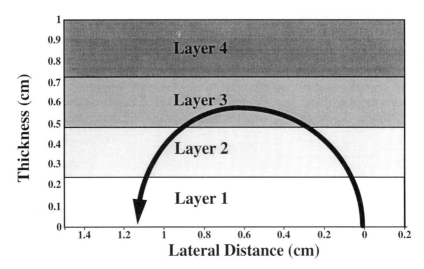

Figure 3 *Schematic of an idealized light path through an multi-layer absorbing/ scattering medium where a source is located at position 0 and a detector is placed some lateral distance away.*

distribution can also be viewed from the perspective of the detector. Instead of a point source of light entering a system, light exits the sample into a point detector situated in the same orientation. Each point now represents the fraction of the total intensity collected by the detector. However, for a given source/detector pair, the fraction collected by the detector will be dependent on the illumination to each spatial position in the sample. To account for the interdependence, the product of the two fractional contributions can be used to estimate the likely distribution through the sample in which an emitted photon will reach the detector.[10]

Examples of the detected photon distributions for two different lateral source and detector separations are shown in Figures 4 and 5. Figure 4 shows this distribution when the source and detector are separated by 6 mm. Each contour in Figure 4 represents a 10% incremental change in the fraction of light traversing a region. The overall distribution of collected light is localized between the source and detector and has a 'banana shaped" appearance. In particular, the region immediately in front of both the source and detector has the highest collected fraction and would therefore be the most sensitive to changes in sample absorption. Regions deeper in the sample contribute less to the collected light fraction. The mean depth from which light is collected is approximately 3 mm.

For the 12 mm separation of the source and detector shown in Figure 5, several differences in the collected light distribution are apparent as compared to results in Figure 4. Overall, the light is collected from a more diffuse region in the sample. There is less definition even at the source and detector positions. Likewise, at the 50% contour, the light extends beyond the 1 cm boundary of the sample. The mean depth of the collected light is approximately 5 mm. Hence, as the detector is moved further away from the source, information regarding deeper layers can be obtained.

An estimate of the surface measurements can be obtained using the results from the Monte Carlo simulation for samples of different crossectional composition. As mentioned in the experimental section, a sample with four distinct absorption layers was simulated in the same way as the sample shown in Figure 3. The remittance as a function of source/detector lateral distance for a series of multi-layer slab compositions is shown in Figure 6. To allow comparisons between the separate simulations, the measured remittance at each lateral position is referenced to the source intensity to obtain the attenuation. The general trend shown in Figure 6 is that absorption deep within a sample has a greater effect on the attenuation made at large lateral spacing of the source and detector as compared to measurements made at small lateral spacings.

To determine if a similar relationship to the Monte Carlo results is found in a real scattering sample, measurements of the experimental phantom described in the methods section were made. The results of a single absorber placed at various depths in a scattering sample are shown in Figure 7. Like the Monte Carlo results, the presence of an absorber has a marked effect on the surface light attenuation. In the experiment, the use of the high concentration dye absorber made the absorption very high and its effect is clearly observed. Because the simulation used relatively low absorbances of 0.17 to 0.70 mm^{-1}, the results are not as pronounced. For the lateral spacing of 5 and 10 mm shown as dashed lines, considerable difference in the effect of the depth position of the absorber is seen. A detector positioned at 5 mm exhibits large changes in attenuation when the absorber is close to the surface (as compared with no absorber) whereas at 10 mm detector spacing, changes in attenuation occur throughout the absorber placement.

6 CONCLUSIONS

The results presented demonstrate the feasibility of depth resolved absorbance determinations based on diffuse reflectance measurements. Both the Monte Carlo results and experimental measurements made in known phantoms suggest that the depth of collected light is highly dependent on the spacing between the source and detector. In

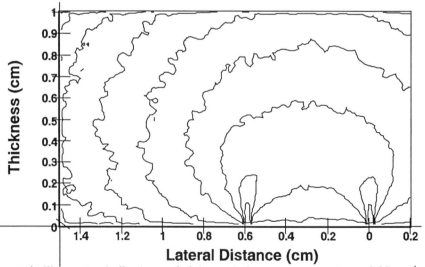

Figure 4 *Illumination/collection probability with 6 mm separation ($\mu_a = 0.35\ mm^{-1}$, μ_s*

particular, as the detector is moved further from the source, information regarding deeper layers in a sample may be obtained. With the information gained through this work, reconstruction methodologies can be developed which will provide a depth resolved absorption measurement. In the future, this technique may provide a novel tool to study spatia of chromogenic constituents in scattering media.

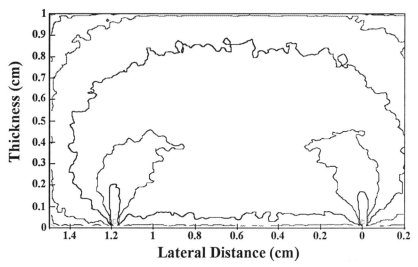

Figure 5 *Illumination/collection probability with 12 mm Separation ($\mu_a = 0.35$ mm^{-1}, $\mu_s = 8$ mm^{-1}, g = 0.97)*

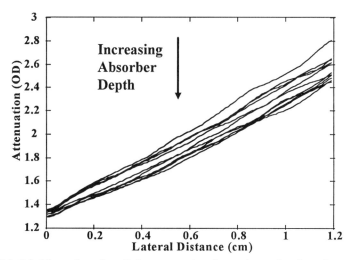

Figure 6 *Modeled lateral surface light attenuation for various absorbers in a slab ($\mu_s = 8$ mm^{-1}, g = 0.97)*

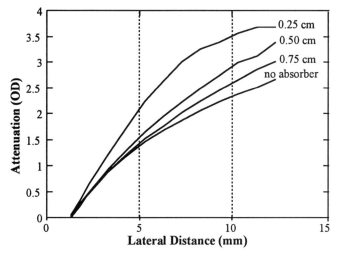

Figure 7 *Experimental lateral surface light attenuation with absorber present*

References

1. D. R. Marble, D. H. Burns, and P. W. Cheong, *Appl. Opt.*, 1994, **33**, 1279.
2. L. S. L. Arakaki and D. H. Burns. *Appl. Spectros.*, 1992, **46** (12), 1919.
3. M. Cope, D. van der Zee, M. Essenpreis, S. R. Arridge, and D. T. Delpy, SPIE Proc., 1991, **1431**, 251.
4. M. Cope and D. T. Delpy, *Med. & Biol. Eng. & Comput.*, 1988, **26**, 289.
5. W. Cui, C. Kumar, and B. Chance, *SPIE*, 1991, **1431**, 180.
6. W. Cheong, S. A. Prahl, and A. J. Welch, *IEEE Journal of Quantum Electronics*, 1990, **26**, 2166.
7. B. C. Wilson and S. L. Jacques, *IEEE Journal of Quantum Electronics*, 1990, **26**, 2185.
8. M. Keijer, S. L. Jacques, S. A. Prahl, and A. J. Welch, *Lasers Surg. Med.*, 1989, **9**, 148.
9. J. C. Hebden and R. A. Kruger, *Med. Phys.*, 1990, **17**(1), 41.
10. J. C. Schotland, J. C. Haselgrove, and J. S. Leigh, *Appl. Opt.*, 1993, **32**(4), 448.

FLUORESCENCE IMAGING OF WHOLE MICROORGANISMS WITH SCIENTIFIC GRADE CCDS

Kenneth D. Hughes, Diana L. Bittner, and Greta A. Olsen
School of Chemistry and Biochemistry
Georgia Institute of Technology
Atlanta, GA 30332

1 INTRODUCTION

The use of microorganisms and small aquatic organisms as monitors of water and sediment quality is increasing and therefore the development of improved methodology and new technologies that facilitate the handling of these organisms is a high priority.[1-3] Methodological changes that have occurred in the last several years have included a concerted effort to monitor changes in organism "health" at the molecular level. This is in stark contrast to assays that simply measure a single data point such as the LC50 (lethal concentration of toxicant where 50% of the organisms die). The ultimate goal in these molecular level assays is to rapidly ascertain the very first sign of stress in an organism exposed to a water- or sediment-based toxicant. One of the first indicators of stress in many organisms is a perturbation in enzyme activity associated with metabolism or detoxification. Perturbations in enzyme activity may be monitored with high sensitivity by utilizing fluorescence techniques.

Monitoring enzyme activity in microorganisms with fluorescence technology involves the use of fluorogenic substrates, which are molecular compounds that exhibit little or no fluorescence until a chemical functionality on this compound is modified by action of an enzyme. These fluorescent analogues of the enzyme's "natural" substrate can be synthesized with common dyes such as rhodamine and fluorescein.

Design of rapid toxicity assays for accessing water and sediment quality demands simple experimental protocols that do not involve complicated fluorogenic substrate delivery mechanisms, time consuming steps for the separation and isolation of the resulting fluorophore, and expensive fluorescence intensity measurement instrumentation. Exposing whole microorganisms to fluorogenic enzyme substrates and monitoring the fluorescent emission from the organism is straightforward and in most cases simply involves introduction of the fluorogenic substrate into the solution containing the organism. After diffusion across the organism's cellular membranes, the fluorogenic substrate is cleaved, yielding the fluorescent dye. Quantitation of the resulting fluorescence is completed by visual notation (+ or -) or by utilizing charge-coupled device detectors and image analysis hardware and software. Some of the variances and limitations associated with these assays include factors related to biology, chemistry, and optical imaging techniques. Improvement in any one of these areas ultimately translates into more sensitive and reliable ecological data. Recently, fluorogenic substrates incorporating micron-diameter polymeric particles have been synthesized and investigated as novel means of quantitating enzyme activity in the digestive tract of microorganisms. Early experiments with these probes indicate that the some of the limitations of using soluble enzyme substrates can be reduced and in some cases eliminated.

2 RESULTS AND DISCUSSION

Design of sensors based on whole microorganisms exposed to fluorogenic substrates (biological sensors) must include the same fundamental characteristics as the design of chemical sensors based upon fluorescence technology. These fundamental characteristics are summarized in Table 1.

Table 1 *Characteristics of Good Biological Sensors*

Simplicity
Reproducibility
Sensitivity
Ecological Relevance
Selectivity
Reasonable Cost

In an effort to address each of these characteristics, a fundamentally different means of exposing microorganisms to fluorogenic substrates has been developed. This new direction involves fabricating fluorogenic enzyme probes based upon the chemical modification of micron-sized spheres. These chemically-modified spheres (CMS) consist of a fluorophore-impregnated particle with a fluorogenic enzyme probe covalently bound to the particle's outer surface. Enzymatic activity occurring at the particle surface produces a fluorophore that emits in the UV/VIS/NIR region of the spectrum. The impregnated fluorophore has a spectral emission also in the UV/VIS/NIR, but separated from that of the fluorescent enzyme probe. The emission of this fluorophore is unaffected by the chemical nature of the particle's environment and serves as a reference signal for the spectroscopic measurement.

Significant advantages are obtained by calculating the ratio of the two fluorophore emissions (enzyme generated and reference) generated by a single excitation wavelength. Variations in the particle or ingestion rate of individual organisms, which determines the amount of enzyme substrate delivered to the digestive tract, and fluctuations in excitation source intensity can be corrected.[4] Fluorogenic enzyme probes based upon micro-diameter particles provide less invasive and more sensitive detection of enzyme activity in microorganisms. This is a direct result of improving the targeting mechanism of the fluorogenic substrate for the digestive tract of the microorganism. The sensitivity of the fluorescence measurement is increased since the enzyme-generated fluorophore is permanently localized at the particle's surface and is unable to leak from the organism. These advantages, as well as others, are discussed in Reference 4.

2.1 Prototype Fluorogenic Substrates

The construction of a prototype enzyme probe is diagrammed in Figure 1. These chemically-modified spheres (CMS) were designed to target esterase-related enzymes and to serve as an initial demonstration of this enzyme probe technology. This esterase probe is composed of the non-fluorescent molecule carboxyfluorescein-diacetate covalently bound to the surface of a one micron diameter latex sphere impregnated with the fluorophore nile red. Coupling this particular enzyme substrate to the particle surface is completed in a single step reaction by utilizing an amine functionality on the particle surface and a succinimidyl ester functionality on the enzyme substrate (carboxyfluorescein-diacetate).

As a result of the high sensitivity afforded by fluorescence assays, significant attention must be directed towards purification of the product. Unreacted reagents and fluorophores (converted substrates) must be removed with several chromatographic steps to assure very low spectroscopic background levels. The need for these traditional

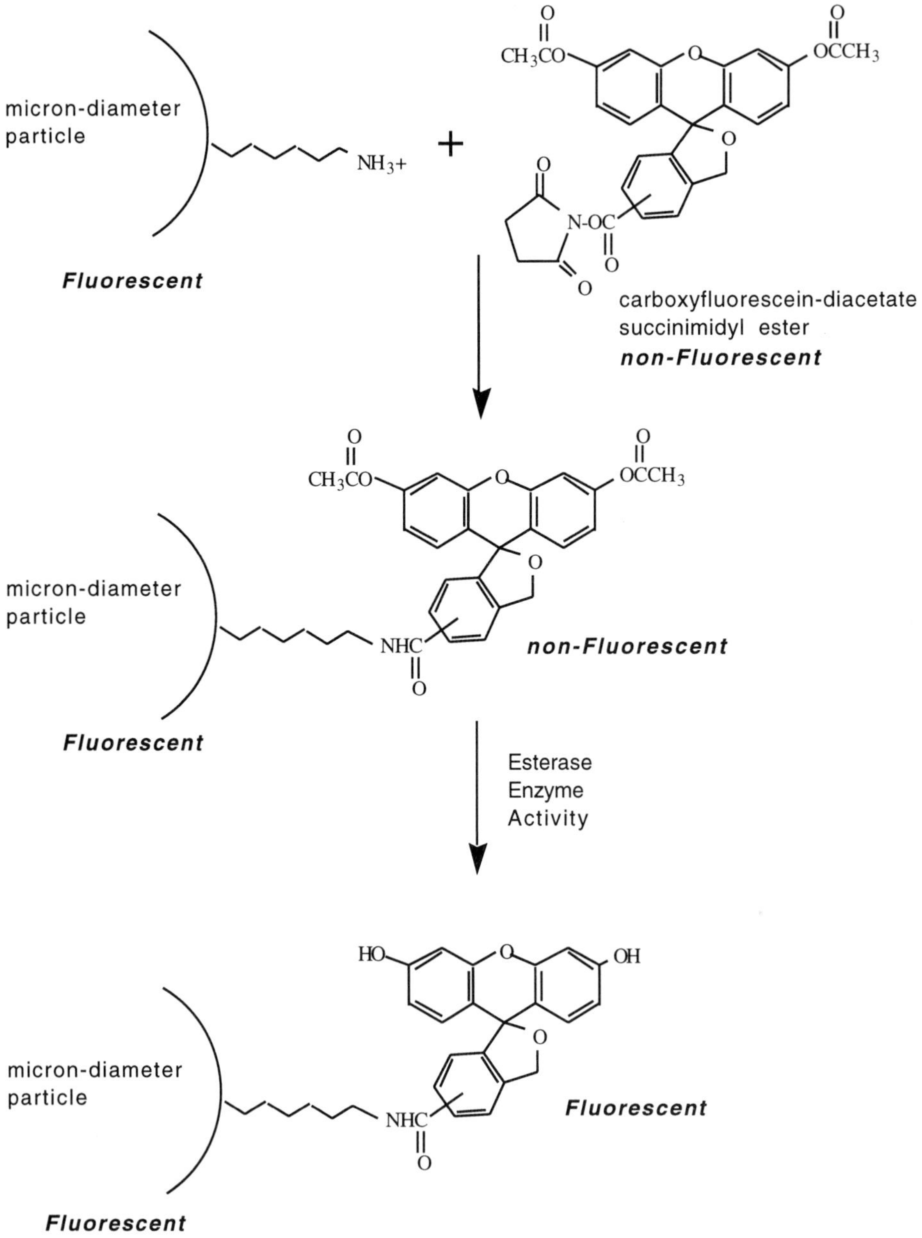

Figure 1 *Design of a prototype probe. These chemically-modified spheres will target esterase-related enzymes. The probe incorporates the non-fluorescent molecule carboxyfluorescein-diacetate covalently bound to the surface of a one micron diameter latex sphere impregnated with the fluorophore nile red.*

chromatographic steps is eliminated when synthesizing enzyme probes based upon micron-sized particles, since the chemical reagents can be removed from the particles by quick dialysis protocols.

The fluorescence signals in this prototype system are generated by exposing the particle and covalently bonded fluorogenic substrate to blue-green radiation (488 nm). The emission from the particle, due to the presence of the reference fluorophore nile red, is observed at 605 nm while enzymatic cleavage of the acetoxymethyl esters in the fluorogenic substrate yields the highly fluorescent fluorescein molecule, which has an emission maximum near 520 nm.

2.2 Initial Organism Exposures

Several different organisms have been exposed to the chemically-modified spheres targeting esterase-related enzymes, including rotifers, larval shrimp and oysters, and single cell organisms such as algae and bacteria. Exposure of organisms to fluorogenic substrates based upon micron-diameter particles is similar to traditional substrates except for the fact that the particle-based probes cannot cross the organism's cellular membranes. Organisms must ingest the particles as they would their natural food sources through filter feeding mechanisms, which provides previously unattainable selectivity to assays based on these chemically modified spheres. By eliminating the fluorescence signal generated by single cell organisms present along with larger microorganisms in heterogeneous solutions, direct analysis of environmental and culture samples can be performed.

One of many fields that could benefit from this new fluorescence technology includes aquaculture.[5-9] Perturbations in enzyme activity and ingestion rate in rotifers (food source for larval organisms in the aquaculture industry) and other small aquatic animals, such as larval shrimp, are sensitive and rapid indicators of environmental stress associated with exposure to organic and inorganic compounds in the organism's environment. This technology could provide an additional avenue towards monitoring the health of a culture by monitoring both water quality (DO, ammonia, etc.) and the organisms digestive or metabolic activity.

Figure 2 provides a white light/fluorescence image of the food shrimp species *Penaeus*, obtained with 20x magnification, long pass 515 nm filter (Schott), 488.0 nm argon-ion radiation (Coherent, Inc.) and a 35 mm camera attached to an Olympus IMT2 Inverted Microscope. The particles ingested by this organism (Figure 1) target esterase enzyme activity. The one-micron-diameter particles are clearly localized in the digestive tract, as noted by the green (fluorescein) fluorescence intensity resulting from esterase activity directed at the particle's surface. Since a long pass wavelength filter was used to capture this image, both nile red and fluorescein signals are present; however, due to the weaker emission of the nile red fluorophore, only fluorescein emission can be seen in the digestive tract of this organism. Signal from nile red is evident in solution around the organism (small spheres) where there is no enzyme activity to generate fluorescein emission.

In this particular application, it is the physical attribute (size) of the probe that facilitates specific targeting of the organism's digestive tract and subsequent excretion of the enzyme probe. In investigating this organism it was observed that approximately three day old larvae would only ingest particles one micron in diameter and smaller. Therefore, it seems possible that cultures could be selectively monitored by age of individual organism. The use of CCD cameras to collect the fluorescence images has allowed quantitation of both particle ingestion rate and enzyme activity. One of the unique features of this new technology is the ability to simultaneously quantitate both the number of particles ingested by the organism and the metabolic activity of the organism. In previous work,[10,11] researchers have screened field sites for toxicity by using soluble enzyme substrates to probe the magnitude of enzyme activity as well as utilizing micron-diameter particles in separate assays to assess ingestion rates.

Improvement in sensitivity of these toxicity assays and decrease in exposure time (fluorogenic substrates) required to obtain a measurable fluorescence signal should be obtainable by using scientific grade CCDs; therefore, these investigations are under way.

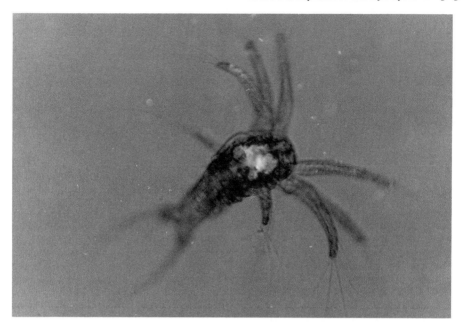

Figure 2   *The food shrimp Penaeus, while in a larval state, was exposed to chemically-
modified spheres targeting esterase enzymes. The green fluorescence is due to
fluorescein emission generated by esterase activity. Signal from nile red is
evident in solution around the organism (small spheres). A 515 nm long pass
filter, 488.0 nm excitation radiation, and 20x Olympus objective were used to
capture the image.*

Significant quantum efficiency in the red region of the spectrum, inherent in these
detectors, will provide sensitive detection of the particle's emission as well as quantitation
of the signal generated by enzyme activity.

ACKNOWLEDGEMENTS

This work has been supported by the Georgia Tech Research Corporation, the Georgia
Tech Institute for Bioscience and Bioengineering, and the U.S. Geological Survey.

References

1. K. R Hayes, W. S. Douglas, Y. Terrell, J Fischer, L. A. Lyons, and L. J. Briggs, *Bull.
Envir. Contam. Toxicol.*, 1993, **51**, 909.
2. C. R. Janssen and G. Persoone, *Env. Tox. Chem.*, 1993, **12**, 711.
3. T. W. Snell and G. Persoone, *Aquatic Toxicology*, 1989, **14**, 81.
4. K. D. Hughes, D. L. Bittner, and G. A. Olsen, *Analytica Chimica Acta: Special Issue*:
1995, **307**, 393.
5. 'Diseases of Cultured Penaeid Shrimp in Asia and the United States', W. Fulks and K.
L. Main (eds.), The Oceanic Institute, Makapuu Point, Hawaii. 1992, p. 392.

6. C. J. Sinderman, 'Diseases of Cultured Penaeid Shrimp in Asia and the United States', W. Fulks and K. L. Main (eds.), The Oceanic Institute, Makapuu Point, Hawaii.,1992, p. 325.

7. B. LeaMaster,. 'Diseases of Cultured Penaeid Shrimp in Asia and the United States', W. Fulks and K. L. Main (eds.), The Oceanic Institute, Makapuu Point, Hawaii., 1992, p. 345.

8. J. A. Brock, 'Diseases of Cultured Penaeid Shrimp in Asia and the United States', W. Fulks and K. L. Main (eds.), The Oceanic Institute, Makapuu Point, Hawaii., 1992, p. 209.

9. D. V. Lightner and R. M. Redman, 'Culture of Marine Shrimp: Principles and Practices', A. W. Fast and L. J. Lester (eds.), Elsevier, New York., 1992, p. 573.

10. S. E. Burbank and T. W. Snell, *Environmental Toxicology and Water Quality: An International Journal*, 1994, **9**, 171.

11. C. M. Juchelka and T. W. Snell, *Arch. Environ. Contam. Toxicol.*, 1994, **26**, 549.

HIGH RESOLUTION X-RAY MICROTOMOGRAPHY SYSTEM USING A DIRECTLY-COUPLED STRUCTURED SCINTILLATOR AND CCD DETECTOR

K. Moon[1], B. R. Dobson[2], R. Bell[3] and N. M. Allinson[4]

[1]   Department of Electronics, University of York, York YO1 5DD
[2]   CCLRC, Daresbury Laboratory, Warrington WA4 4AD
[3]   EEV Ltd, Chelmsford, CM1 2QU
[4]   Department of Electrical Engineering and Electronics, UMIST, Manchester M60 1QD

1 INTRODUCTION

X-ray microtomography (XMT) is a relatively new exploratory technique that can provide detailed two- and three-dimensional non-destructive internal images of materials and organic samples. The technique relies on measuring the x-ray absorption for numerous projections through a sample. Computer algorithms can then be used to reconstruct the original absorption densities within the volumetric space of the sample. Its essential features are those of the better known medical computer tomography (CT). While CT scanners use conventional Bremsstrahlung x-ray sources, with a fan-shaped beam intercepted by curved arrays of photomultiplier tubes coupled to scintillator crystals; XMT systems take advantages of the unique properties of synchrotron radiation (SR). These advantages include not only the source's high intensity but its natural vertical collimation and tuneability. For example, the horizontal source size at the CCLRC synchrotron source at Daresbury Laboratory, Cheshire, UK, the horizontal source size is 1.3 mm. At a distance of 30 m from a tangent point on the ring, the source subtends an angle of approximately 40 μrad. If a detector with linear pixel dimensions of 10 μm is placed about 20 cm behind the sample, then the resolution is determined solely by the detector. The ability to tune to particular wavelengths means that data collection can be optimized to suit the densities and composition of individual samples, or collection can be made at different wavelength so identifying the spatial arrangement of individual elements.

Several XMT systems have been constructed at SR installations with limiting imaging volumes ranging from 50 μm down to 1 μm cubes (voxels).[1,2,3] These small voxels should be compared with typical CT values of 2 – 5 mm voxels. The high resolution of XMT can be exploited for the microanalysis of geological samples (e.g., the granular structure of sandstones for the potential recovery of petroleum reserves), material samples (e.g., the microstructure of composite materials or the distribution of catalysts within their support matrix), and biological samples (e.g., the distribution of calcium and phosphorus in bones can provide clues as to the effects of disease on bone strength, etc.).

Possible detector arrangements are shown diagrammatically in Figure 1. For the finest spatial resolution then a single pin-hole detector is the most suitable, but this results in very inefficient data collection. A more efficient approach is to employ one-dimensional array detectors for single cross-sectional reconstructions or two-dimensional arrays for full volumetric reconstructions. The typical CCD pixel dimensions are 6 – 25 μm, which places an upper limit on resolution unless magnifying optics are inserted between the primary scintillating screen and the CCD imager. However, the light transmission of a lens system is approximately given by $(NA)^2$, where NA is the system's numerical aperture. Typical efficiencies for high-quality microscope objectives are only about 5%. Such a large reduction in the detected CCD signal will result in excessive data collection times if reconstruction accuracy is to be maintained. In practice, the limiting resolution is often set by the scintillator, due to the scattering of the scintillation photons. There are conflicting requirements of ensuring high x-ray absorption (i.e., thick scintillator) and low optical scattering (i.e., thin scintillator). Custom cellular scintillator screens have been developed for XMT systems,[4] which consist of differentially etched fibreoptic plates where the missing core volumes are filled with scintillating material.

This paper discusses an alternative approach to XMT system design where the emphasis is placed on efficient data collection and yet retain high spatial resolution. The work described here should be viewed as an initial feasibility study to assess the potential of this approach. The design employs 1:1 optics based on fibreoptic studs directly coupled to a CCD imager, and a structured scintillator evaporated *in situ* on a stud. Also of interest is the use of dither clocking techniques to reduce the thermal noise of a standard CCD without the need for excessive cryogenic cooling. Specimen results are also given which provide an exciting glimpse of the potential of XMT.

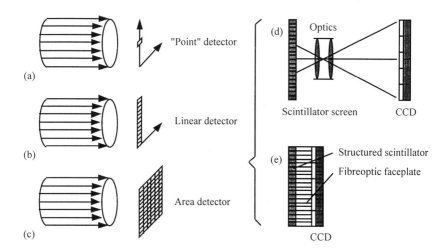

Figure 1 *Microtomography detector arrangements. Option (e) is employed in the system described here*

2 TOMOGRAPHIC IMAGING REQUIREMENTS

Transmission tomography creates cross-sectional images of the internal structures for a sample by processing numerous projection images produced by the penetrative x-ray beam. The following will consider 2-D reconstruction of a single cross-section, as a volumetric reconstruction can be thought of as a simple stack of these slices. The intensities of the incident, I_o, and transmitted, I_t, beams at a beam direction, θ, and impact parameter, t, for a particular ray path are related by the optical depth function, $p(\theta, t)$.

$$p(\theta, t) = \ln\left(I_o / I_t \right) = \int_s f(x, y)\, ds \tag{1}$$

where $f(x, y)$ is the linear x-ray attenuation coefficient and s is the corresponding ray path length. Figure 2 shows the relationship between the sample axes and those of the source-detector frame.

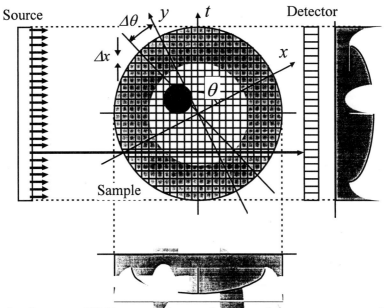

Figure 2   *Geometry of XMT system indicating detector and reconstruction co-ordinate frames*

2.1   Reconstruction Accuracy

Reconstruction is achievable using the following Fourier transform pair.

$$F'(u, v) = \iint f(x, y)\, e^{-2\pi j(ux + vy)}\, dx\, dy \tag{2}$$
$$f'(x, y) = \iint F(u, v)\, e^{2\pi j(ux + vy)}\, du\, dv$$

Hence, relating this pair to the recorded projection data

$$P'(\theta, \omega) = \int p(\theta, t) e^{-2\pi \omega t}\, dt = F'(u, v) \tag{3}$$

since $u = \omega \cos\theta$ and $v = \omega \sin\theta$. If the reconstruction is bounded to a finite spatial resolution, that is $f(x, y)$ is a band-limited function with $F'(|u|, |v|) = 0$ for $|u| \geq u_c$ and $|v| \geq v_c$, where u_c and v_c are the spatial cut-off frequencies (assumed to be the same in the following approximate analysis). To match this precision of possible reconstruction then $2 v_c \Delta t \leq 1$, where Δt is the impact parameter increment.[5] This expression is simply a restatement of Nyquist's sampling theorem. This band-limitation is, normally, imposed by the detector's spatial resolution (i.e., Δt). To reconstruct to a linear spatial resolution of 20 μm then the detector element pitch must be at least 10 μm. To provide uniform spatial resolution in the *x-y* plane, then simple geometric considerations imply that

$$\Delta\theta \leq \frac{2\Delta t}{D} \tag{4}$$

where D is the nominal diameter of the illuminated sample. Hence, the number of equispaced projection images, M, that must be obtained for $\theta = 0, \ldots, \pi$, is

$$M \geq \frac{\pi D}{2\Delta t} \tag{5}$$

The noise amplification, g, inherent in tomographic reconstruction has been shown to scale as[1]

$$g^2 = \left(\frac{\sigma_f p}{\sigma_p f}\right)^2 \approx \frac{D\Delta t}{\Delta x^2} \approx N \tag{6}$$

where N is the number of vertical pixels in the detector. The right-hand approximate equality holds when $\Delta x \approx \Delta t$. For example, for a relative accuracy in the reconstructed tomograph of 1%, then with $N = 160$ (as in our case), the relative error, σ_p/p, of the projection image must be better than 0.08%. The minimum number of incident photons per projection image, N_0, required to ensure a reconstructed image with a relative accuracy of σ_f/f is :

$$N_o = \frac{g^2 \exp(\tau)}{\left(\tau\sigma_f \Big/ f\right)^2} \tag{7}$$

where $\tau (= fD)$ is the average attenuation through the sample.[6] This function is a minimum when $\tau = 2$. The physical reason for this minimum is that the quantity used in the reconstruction is the beam attenuation – a low attenuation means poor accuracy; a high attenuation results in a small, and hence inaccurate, measured transmitted beam. At this optimum condition, where $\tau = 2$, specified reconstruction accuracies of 1% and 5%, and a 160 x 160 cross-section, require an incident flux of $N_o \approx 3.10^6$ and $N_o \approx 2.10^5$ detected photons per detector pixel respectively.

2.2   Detector Requirements

A detector system's detective quantum efficiency, *DQE*, is given by

$$DQE = \frac{\left(S_o \Big/ \sigma_o\right)^2}{\left(S_i \Big/ \sigma_i\right)^2} = \frac{\left(p \Big/ \sigma_p\right)}{N'_o} \tag{8}$$

where $S_i \Big/ \sigma_i$ and $S_o \Big/ \sigma_o$ are the input and output signal-to-noise ratios, and N'_o is the required incident flux to yield a specified reconstruction accuracy. For an ideal detector, $DQE = 1$ and $N'_o = N_o$; for a practical non-intensified CCD system, the *DQE* can be expressed as[7]

$$DQE = \frac{1}{\dfrac{1}{\varepsilon_s}\left(1 + \dfrac{1}{g_s T_f \eta_{ccd}}\right) + \sigma_{eff}} \tag{9}$$

where
       ε_s scintillator absorption
       g_s scintillator gain (detectable visible photons per interacting x-ray photon)
       T_f transmission efficiency of optical relay (lens or fibreoptics)
       η_{ccd} CCD quantum efficiency
       σ_{eff} effective additional "electronic" noise.

In minimizing the data collection times (i.e., as $N'_o \rightarrow N_o$) then the importance of high scintillator absorption and efficient optical coupling can be appreciated from (9). Hence, our exploitation of non-magnifying "optics", optimized optical transmission paths and high-absordance structured scintillator. The other factors in this expression for DQE are relatively consistent with other system designs. Also the previous section highlighted the need for high full-well capacities in the CCD as the detected flux per pixel needs to be high in order to ensure good reconstruction accuracy.

3 EXPERIMENTAL ARRANGEMENT

The experiments were conducted using the Synchrotron Radiation Source (SRS) at CCLRC Daresbury Laboratory on Station 7.4, which is about 40 m from a ring bending magnet on the SRS. This distance ensures a very well collimated beam. The general layout of the beamline components is shown in figure 3.

3.1 Beamline Components

A set of entrance slits coarsely defines the beam aperture to the tuneable Si-220 channel-cut monochromator. A set of exit slits more closely defines the aperture of the nominal 8 keV monochromatic beam incident at the sample. With this arrangement, a relatively uniform illumination could be achieved over a 8 mm (H) x 9 mm (V) aperture. Unfortunately, a spurious diffraction spot (due a monochromator crystal blemish) limited the usable aperture to 8 mm (H) x 3.6 (V) mm. Ion chamber measurements indicated an incident flux of approximately 8,000 x-ray photons/CCD pixel/s. A motorised rotary shutter permits accurate shuttering of the detector during readout. The sample is mounted on a precision rotary table. Rotation accuracy is better than one second of arc; and the sample can be positioned horizontally, using a translation stage, with a repeatability of ±1 μm. This stage can also be used to fully withdraw the sample from the beam in order to permit regular incident beam calibrations. The detector head is mounted on a multi-axis stage (vertical, transverse and tilt). Accurate alignment and re-positioning of all beamline components is essential in microtomographic studies. Pixel (voxel) misalignment/ repeatability must be less than 0.5 pixel (voxel) for accurate reconstruction.

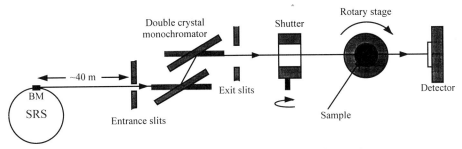

Figure 3   *Experimental arrangement – SRS: Synchrotron Radiation Source; BM: Bending magnet*

3.2 Acquisition System

The acquisition system was originally developed for high-speed x-ray crystallography, notably Laue diffraction studies[8]. However, it was designed as a flexible, general-purpose diagnostic instrument with a wide variety of interfacing and operating modes. The remote camera head, for this work, consists of a front-lit EEV CCD02 device (576 x 384 pixels on 22.5 μm x 22.5 μm pitch) on to which is directly mounted a 5 mm thick fibreoptic stud. The limited region of uniform illumination at the sample restricts the usable aperture to 354 x 160 pixels. Oil-mounted on the front surface of this stud is a further 1:1 fibreoptic stud with an *in situ* structured CsI scintillator. The head also contains a two-stage thermoelectric cooling module. Overall conversion gain is Å10 equivalent e$^-$/x-ray photon. For a pixel rate of 2 MHz, the system rms noise floor is Å 80 e$^-$ (at 295 K) and a saturation charge of Å 4.10^5 e$^-$ per CCD pixel.

3.3 Structured Scintillator

Over a dozen materials have been regularly employed as inorganic x-ray scintillators – the most widespread being the powder phosphors Gd_2O_2S:Tb ("Gadox") and Y_2O_2S. We have concentrated on the development of scintillators where the scintillating material is grown *in situ* on a supporting substrate.[9] CsI has long been a popular material, especially in bulk form, but it can be evaporated onto to glass and other substrates where it forms thin needle-shaped crystallites. Doping CsI with Tl will significantly increase the conversion efficiency as well as provide an emission spectrum that closely matches the spectral sensitivity of a CCD (λ_{max} Å 540 nm). It also demonstrates a high x-ray stopping power, fast response (no long lifetime emission components above the 0.01% peak emission level, with careful choice of source material) and high detective quantum efficiency, DQE, up to E_λ Å 40 keV (limited by the generation of fluorescent photons above the cesium and iodine K absorption edges).

One of the most important parameters for any scintillator is its spatial resolution. This is particularly true for microtomography applications. The overall system's spatial resolution, and low-level DQE, performance is usually determined by the light scattering properties of the primary converter. To preserve the inherent resolution of the fibreoptic elements, after suitable treatment the front surface of the fibreoptic stub is differentially etched to leave the cores of the individual fibres proud. On to this "bed of nails", the CsI is vapor deposited to form crystallites about 1 – 3 μm in diameter. The spatial resolution, expressed in terms of the point spread function, PSF, shows about a three-fold improvement over using an untreated fibreoptic. In fact, the measured x-ray response PSF approaches that of the visible response PSF (i.e., limited by the fibreoptic components and optical interfaces). The PSF for the full microtomography acquisition system is shown in Figure 4 (PSF_{FWHM} = 34 μm; $PSF_{1\%}$ = 130 μm). The use of structured scintillators also yields a significant improvement in the conversion efficiency due to the enhanced optical coupling between the fibre cores and the scintillator.

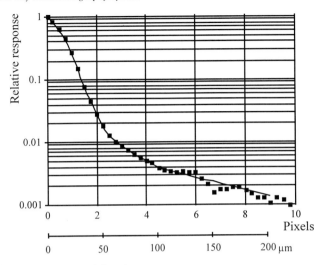

Figure 4   *System point spread function*

3.4   Dark Current Reduction

As the frame integration periods are typically over one second and CCD cooling is limited to about 8 C (in order to guarantee a safety margin before water condensation occurs), dither clocking is employed to reduce dark current to an acceptable level. During the integration period, the CCD is operated with its substrate bias high enough to invert the silicon surface under a gate electrode at the clock-low level; and periodically the clock-high gate electrode is alternated within each CCD pixel. When the surface is depleted, the thermal generation current due to interface states will be at a maximum. This current is usually the major contribution to dark current in buried-channel CCDs. The steady-state surface dark current, $J_s(\infty)$, is

$$J_s(\infty) = qn_i s_o \qquad (10)$$

where n_i is the intrinsic carrier concentration and s_o is the surface generation velocity. If the surface is inverted, the current generation is suppressed. Due to the dynamics of this generation, the rate of generation can be significantly reduced. Burke and Gajar[10] have shown how the approximately exponential relationship between this switching period and $J_s(t)$ can be exploited to provide very low dark current operation for standard CCDs without resorting to excessive cryogenic cooling.

The recovery of surface dark current following inversion exhibits a time constant, τ, given by

$$\tau = \frac{1}{n_i v_{th} \sigma} \qquad (11)$$

where v_{th} is the carrier thermal velocity and σ is the trap cross-section (i.e., $\sigma = \sqrt{\sigma_n \sigma_p}$, where σ_n and σ_p are the electron and hole capture cross-sections respectively). The normalised variation in surface dark current with recovery time is

$$\frac{J_s(t)}{J_s(\infty)} = \frac{2}{\pi} \int_0^\infty \frac{1 - \exp\left[-\frac{t}{\tau}\left(x + x^{-1}\right)\right]}{1 + x^2} dx \tag{12}$$

where $x = \sqrt{\sigma_n / \sigma_p} \, \exp\left\{\left(E - E_i\right)/kT\right\}$. For a dither period T, the mean normalized dark charge collected in a CCD element is

$$\frac{1}{T} \int_0^T \left\{ \frac{J_s(t)}{J_s(\infty)} \right\} dt \tag{13}$$

The time constant, τ, is a strong function of device temperature due to its direct dependence on n_i – being 11s at -40 C, 15 ms at 20 C and 180 µs at 80 C. The theoretical reduction of surface dark current using dither clocking is shown in Figure 5. Extensive measurements on several devices and under a variety of operating conditions suggest that these reductions can be achieved. Slight deviations between experimental and theoretical values are probably due to the small dark current component due to bulk thermal generation and inter-electrode fringing effects (which result in a small region of the silicon surface always being in depletion). This second effect could be eliminated by

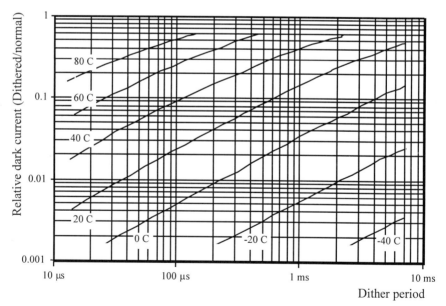

Figure 5 *Relative surface dark current as a function of dither period and device temperature*

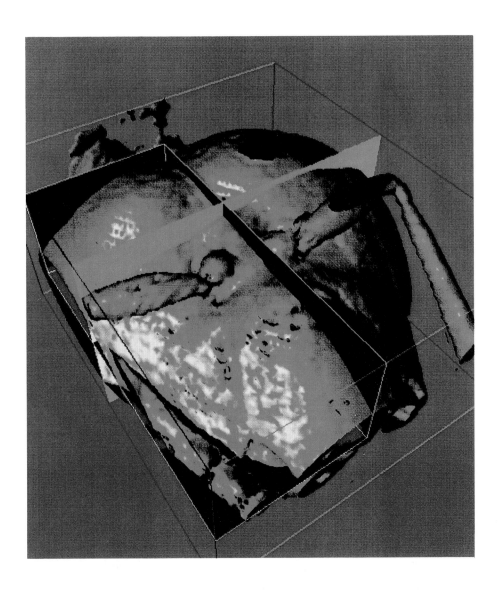

Figure 6 *Tomographi volumetric reconstruction of a worker bee's (Apis mellifera) head*

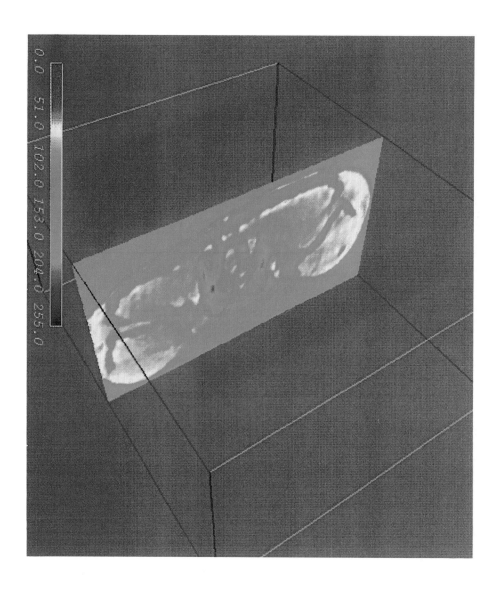

Figure 7 Transverse cross-section through a bee's head

more complex clocking arrangements other than the simple sequencing between adjacent phases as reported here. For the microtomographic study, simple dither clocking only during the integration period gave an effective device operating temperature of -20 C (for a 1 ms dither period).

4 TOMOGRAPHIC RESULTS

4.1 Data Collection

For most experiments, 180 individual projections are recorded at $\Delta\theta = 1°$ intervals. Due to temporal drift in both the incident beam, $I_o(x,y)$, and the system dark current, $I_d(x,y)$, data is acquired in blocks of ten projections with calibration measurements of $I_o(x,y)$ and $I_d(x,y)$ recorded either side of a block. The individual frame integration time remained constant, namely 1.2 s, and each projection consisted of the summation of 5 to 30 frames depending on the sample density and desired reconstruction accuracy. Total time to record a complete data set is between 20 min and 2 hr.

The recorded projection data is corrected, on a pixel-by-pixel basis, for non-uniformities in the white and dark field illuminations, and it is also normalised prior to disc storage. The stored projection data is given by

$$I_n^k(x,y) = \frac{\left(I_t^k(x,y) - I_d(x,y)\right)}{I_c^k(x,y)} \cdot \text{Scale} + \text{Offset} \, , \text{ for } k = 1, 2, ..., K \tag{14}$$

where the interpolated incident beam intensity, $I_c^k(x,y)$, is given by

$$I_c^k(x,y) = \frac{k\left(I_{os}(x,y) - I_{oe}(x,y)\right)}{K} + \left(I_{oe}(x,y) - I_d(x,y)\right) \tag{15}$$

and $I_{os}(x,y)$ and $I_{oe}(x,y)$ are respectively the incident beam calibration frames taken at the start and end of the block (of size K). Even with a stable source such as the SRS, it was found that the spatial drift in $I_o(x,y)$ could be up to 50 μm (i.e., 2 pixels) over a complete experimental run. This drift is observable as additional high frequency noise in the recorded data. Collecting projection data in blocks of ten, with corresponding calibration frames, reduced any temporal or spatial errors to a satisfactory level.

4.2 Specimen Images

The reconstruction algorithms employ the conventional back-projection methods that have become the standard in many CT applications, as they replace the direct Fourier reconstruction (which suffers from a need to perform awkward subsidiary calculations) by numerical operations on the projection data in the object plane. Though the reconstruction software was publicly available, and a commercial visualization package was used, considerable effort was necessary to produce the quality of the results given here.

Figure 6 is the volumetric reconstruction of the head of a worker honeybee (*Apis mellifera*), with a pseudo-coloured surface The data files are in excess of 20 Mbytes and a reconstruction volume of approximately 8 mm x 2.5 mm x 2.5 mm. The reconstruction box has truncated parts of this head, but it is possible to identify the antenna, behind which are the rounded domes of the two compound eyes and in front of the antenna are parts of the bee's mouth-parts. The transverse cross-section as indicated the pale green plane behind the antenna is shown in Figure 7. In this section, the thin exo-skeleton is visible along the top of the section, on either side are sections through the eye structures with the optic nerve extending back towards the brain-stem.

5 CONCLUSIONS

Our study has clearly demonstrated the feasibility of this approach for constructing a high resolution XMT system with the operational advantages of experimental compactness, operational simplicity and high efficiency. Further developments will include transferring the system to a higher intensity beamline and the possible introduction of a second monochromator between the sample and the detector (in order to reduce the effects of *crossfire signals* due to scatter). These improvements, together with modifications to the detector system (e.g., smaller CCD pixels and scintillator improvements) should result in data collection time reductions of over twenty. The use of clock dithering techniques for reduced dark current production would warrant further study.

References

1. B. P. Flannery, H. W. Deckman, W. G. Roberge and K. L. D'Amico, *Science*, 1987, **237**, 1439.
2. C. Tuniz and R. Devoti, *Nucl. Instr. and Meth.*, 1990, **B50**, 338.
3. K. L. D'Amico, J. H. Dunsmuir, S. R. Ferguson, B. P. Flannery and H. W. Deckman, *Rev. Sci. Instrum.*, 1992, **63**, 574.
4. H. W. Deckman, J. H. Dunsmuir and S. M. Gruner, *J. Vac. Sci. Technol.*, 1989, **B7**, 1832.
5. R. M. Mersereau and A. V. Oppenheim, Proc. IEEE, 1974, **62**, 1319.
6. L. Grodzins, *Nucl. Instr. and Meth.*, 1983, **206**, 541.
7. N. M. Allinson, *J. Synch. Rad.*, 1994, **1**, 54.
8. K. J. Moon, N. M. Allinson and J. R. Helliwell, *Nucl. Instr. and Meth.*, 1994, **A348**, 631.
9. C. M. Castelli and N. M. Allinson, *J. X-ray Sci. and Techn.*, 1995, **5**, 207.
10. B. E. Burke and S. A. Gajar, IEEE Trans. Elec. Dev., 1991, **38**, 285.

CHARGE INJECTION DEVICE ABSOLUTE QUANTUM EFFICIENCY DETERMINATION IN THE X-UV RANGE (1-6 KEV) AND DEPLETION DEPTH EVALUATION

Q. HANLEY and M.B. DENTON
Dept. of Chemistry
University of Arizona
TUCSON AZ 86721 (USA)

E. JOURDAIN[+,++], J.F. HOCHEDEZ[++], and P. DHEZ[+,++]
[+] L.S.A.I. and [++] I.A.S.
Université Paris Sud
91405 ORSAY Cedex (France)

1 INTRODUCTION

Charge Injection Devices (CIDs) are, like the more widely known Charge Coupled Devices (CCDs), members of the Charge Transfer Device[1,2] (CTD) family of optical imaging detectors. CIDs have the capability of performing high speed single pixel address, random access integration, real time evaluation of image quality, and non-destructive readout during image acquisition. Due to its unique combination of readout modes and signal integration possibilities, CIDs open a new area in spectroscopy and imaging. While the characteristics of CIDs in the optical photon range have been published, little work has been done to date on the efficiency of CIDs in the far UV and soft X-ray range. The purpose of this paper is to present X-UV Quantum Efficiency (QE) measurements achieved on a CID (CID 17PPRA made by CIDTEC) initially designed for sensing visible light.

This study will be presented in four parts. First, the interaction of photons with Si internal photo effect devices will be summarized. This section will focus on the physical properties of Si that allow CIDs to be used as detectors for the X-UV. In a second part, measurements obtained at the SACO synchrotron facility of LURE are presented. Third, a simple model is compared to the measured results in order to explain the main features of the QE curve and to evaluate depletion depth and oxide layer thickness. Finally, conclusions will be reached with regard to the utility of the devices as detectors in the X-UV range and suggestions made for future optimization of CIDs for this purpose.

These measurements represent the first evaluation of depletion layer thickness in CIDs and also the first reported measurements of the variation in response of a CID with the energy of incident photons.

2 SILICON IMAGING DEVICE IN THE VISIBLE AND X-UV RANGES

The two most important parameters controlling the X-UV detection efficiency in CTDs are the oxide over layer thickness and the depletion depth. Figure 1 is a plot of the pen-

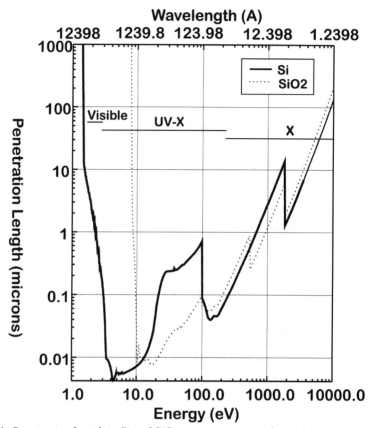

Figure 1 *Penetration length in Si and SiO2 versus energy and wavelength*

etration lengths[3] of photons having energies from 1-10000 eV in Si and SiO_2. The penetration length corresponds to the depth at which 1/e of the incident photons are absorbed (according to the Beer-Lambert law). From such curves it is easily seen that CTDs, despite being designed for visible light, might be useful over a limited region of the X-UV range. The main drawback to the use of devices without specialized engineering for the X-UV is that the SiO_2 layer is very absorbing outside the visible range. In the devices used here, this SiO_2 layer formed a "dead" layer on the order of one micron.

3 EXPERIMENTAL QE MEASUREMENTS FROM 1.7 TO 6 KEV.

A CID 17PPRA device was mounted in a dewer providing liquid nitrogen cooling. Radiation was generated by the LURE Synchrotron Radiation source and photons of appropriate energy were selected using a double crystal monochromator.[4] The photon flux was measured using a calibrated proportional counter as a reference photon detector.

The CID 17PPRA device is a pre-amp per row architecture device fabricated on 5Ω-cm[-1] silicon. The epitaxial layer of the devices varies from 15-38 microns, depending on the particular device. The CID 17PPRA is a 256 X 388 pixel device in which each pixel is 28 X 24 microns.

The QE of the CID 17PPRA was calculated from the integrated signal in the device using Equation 3.1:

$$QE = \frac{S \times 3.65 \times G}{nbph \times E} \qquad (1)$$

In equation 3.1:

- S is the signal in the CID summed over all irradiated pixels (after a dark image subtraction has removed the dark current and the pixel non-uniformity) expressed in ADU (Arbitrary Digit Unit)

- *nbph* is the number of photons reaching the CID based on calibration of the beam with the proportional counter,

- E is the photon energy,

- 3.65 is the energy in eV of an electron-hole pair creation,[5]

- G is the system gain[6] of the camera at which the images have been generated expressed in carriers/ADU.

3.1 Experimental results

Figure 2 shows the absolute QE of the CID 17PPRA over the range from 1.7-6.0 keV. The CID QE curve exhibits two regions of interest. First, the QE varies rapidly in the range near the Silicon K edge (1750 eV). Second, the QE decreases smoothly with energy in the higher energy region. This latter case is due to the penetration length being comparable to, or larger than, the expected depletion depth.

For contrast, Figure 3 shows a Si transmission curve obtained with the same monochromator plotted over the same range as the QE measurements made with the CID. This transmission curve is consistent with that expected for a given energy and the silicon edge features of the CID.

The QE variation around the silicon K edge and the decrease in QE with energy was then selected for more detailed study. Analysis of the latter region indicates the energy at which the depletion depth of the CID becomes smaller than the X-ray penetration depth in silicon. Analysis of these regions gives access to the CID depletion depth.

4 DATA INTERPRETATION.

It is possible to model the two segments of the data presented in the preceding section using a "crude" model. Application of the model to the data allows the depletion depth of the device and the thickness of the SiO_2 to be estimated. The simplest model explaining

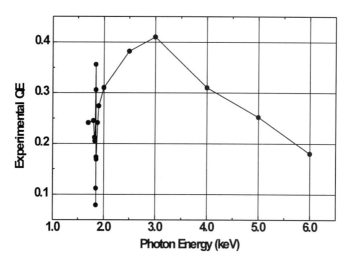

Figure 2 *Absolute QE measurement from 1.7 keV to 6.0 keV*

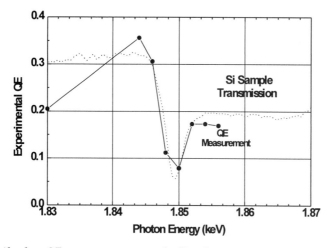

Figure 3 *Absolute QE measurement over the Si$_k$ edge*

the main observed features rests on two assumptions. First, a single absorbing SiO$_2$ over layer is assumed to account for all the photons absorbed before they reach the active silicon region where photoelectrons are efficiently collected. Second, a fully efficient silicon layer having a thickness equal to the depletion depth is assumed to account for all the signal generated in the device.

Figure 4 shows computed QE[7,8] curves generated using this model. For these curves the SiO$_2$ dead layer was set to 2 μm, and the thickness of the active silicon layer was

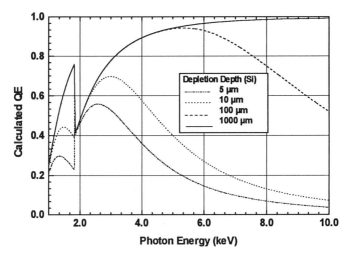

Figure 4 *Energy vs quantum efficiency for various depletion depths using the "crude" model and an assumed SiO₂ upper layer of 2 μm thickness*

allowed to vary from 5-1000 microns. It should be noted that 1000 microns is much greater than X-UV penetration length. Therefore, no decrease of QE versus energy is seen.

Two points may be noted from this curve. First, the energy at which a given QE curve departs from the 1 mm curve depends heavily on the depletion depth of the device. Second, the QE amplitude variation and the shape of the curve in the region of the Si_K edge also depends on the depletion depth.

Figure 5 shows the results of different fits of our "crude" model to the data. Data at high energy, above the Si_k edge, allow both the thickness of the SiO_2 dead layer and the CID depletion depth to be estimated. To check the robustness of the model, a series of fits were made selecting data over different energy ranges, enabling evaluation of the perturbation introduced by the rapidly changing QE in the vicinity of the Si_k edge.

All fits indicated a CID 17PPRA depletion depth of about 5μm. It should be noted that the "crude" model neglects phenomena at the Si-SiO₂ interface and does not account for collection of charges generated outside the depletion region.

5 CONCLUSION

QE measurements in the X-UV have allowed an estimate of the depletion depth of the CID 17PPRA to be estimated. The value of 5 microns measured here is consistent with values expected for a device fabricated on relatively low resistivity silicon. These measurements can bring important information on the CID structures (such as depletion depth and upper layer). As indicated above and clearly shown in Figure 1, the SiO₂ layer should be made as thin as possible. The thickness of the SiO₂ is the primary determinant

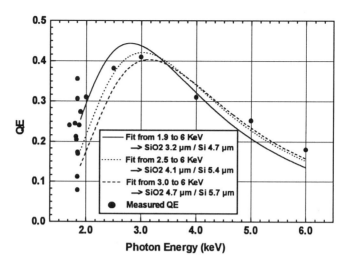

Figure 5 *Fits of the data with the "crude" model*

of quantum efficiency below the Si_K edge. Therefore, as has been done with CCDs, construction of a CID with a thin SiO_2 layer is of interest. For the energy range above 3 keV, the best way to enhance quantum efficiency is to increase the depletion depth. This enhancement may be obtained by fabrication of CIDs on high resistivity silicon. It is also of interest to perform the same measurements at lower energies to explore the aluminum electrode structures and possibly to more precisely account for the different layers in our model. The "crude" model described here does not take into account the aluminum strap connecting the row electrodes or the nitride passivating layer.

ACKNOWLEDGMENTS

Among the colleagues who helped us to achieve this work, we would like to specially thank:

The Bruyères le Chatel B3/CEA group for the help using the Synchrotron Rayonnement source monochromator and obtaining absolute flux measurements.

C. Bizeuil, G. Cauchon, M. Idir, G. Souiller, and P. Stemmler.

M. Naud for software development and electronic debugging.

References

1. J. V. Sweedler, R. B. Bilhorn, P. M. Epperson, G. R. Sims, and M. B. Denton, *Anal. Chem.*, 1988, **60**, 282A.
2. R. B. Bilhorn, J. V. Sweedler, P. M. Epperson, and M. B. Denton, *Appl. Spec.*, 1987, **41**, 1114.

3. E. Palik, 'Handbook of optical constants of solids,' Academic Press 1985.
4. M. Idir, A. Mirone, G. Soullie, Ph. Guerin, Fr. Ladan, and P. Dhez, *Optics Communications*, 1995, **119**, 633.
5. O. Christensen, *J. Apl. Phys.*, 1976, **47**, 689.
6. Q. Hanley, *Seminar at LURE*, January 1995.
7. B. L. Henke, 'Low energy X-ray interactions: photoionization, scattering, specular and Bragg reflection,' in D.T. Attwood and B.L. Henke, eds., Low Energy X-ray Diagnostics (Am. Inst. Phys. Conf. Proc. 75, New York, 1981), 146-155.
8. B.L. Henke, E.M. Gullikson, and J.C. Davis, *X-ray interactions: photoabsorption, scattering, transmission, and reflection at E=50-30,000 eV, Z=1-92*, Atomic Data and Nuclear Data Tables, 181-342, 1993.

HYPERSPECTRAL IMAGE ANALYSIS WITH HYBRID NEURAL NETS

J. M. Lerner, T. Lu, and P. Shnitser

Physical Optics Corporation
20600 Gramercy Place
Torrance, CA 90501 USA

1 INTRODUCTION

The ability to acquire a complete image with an array detector at a large number of wavelengths offers a powerful means of resolving low contrast objects by analyzing their spectral information. Hitherto, the penalty has been the time required to process the vast amount of data generated, and the associated problems of data storage. This paper describes a method of rapidly determining the characteristic spectral features of an object of interest, its background, and other objects. Images are transmitted through an acousto-optic tunable filter controlled by either of two neural nets, one unsupervised and the other supervised. Following neural net training, the AOTF can be programmed to transmit multiple selected wavelengths that form fingerprints of objects of interest. A single frame may, therefore, be all that is necessary for full object identification.

2 HYPERSPECTRAL IMAGING

A *Hyperspectral Image* (HSI) is acquired by capturing an image at multiple wavelengths, usually with a matrix array such as a CCD. Each pixel then contains a complete spectrum limited only by the number of wavelengths selected. An HSI may have between 100 and 200 spectral bands per pixel and a *multispectral* image between 10 and 20 spectral bands per pixel.[1]

Traditional methods of acquiring spectral images of remote objects, in the simplest cases, use filters in front of a camera, implicitly assuming that the wavelengths selected contain the necessary information for identification.[2-4] For more specific wavelength characterization, an image captured by a telescope or microscope is moved across the entrance slit of an imaging spectrometer with a "push broom" action and the final image reassembled by software. The entire image of the entrance slit, and its associated image slice of the remote object, is recorded at all wavelengths simultaneously. In this paper an acousto-optic tunable filter (AOTF) is employed that can be driven to select any wavelength within a specified range. The AOTF enables a user to acquire the entire image simultaneously at one wavelength and then, as the AOTF is scanned, the image is recaptured for as many wavelengths as may be required. A typical AOTF operating geometry is shown in Figure 1.

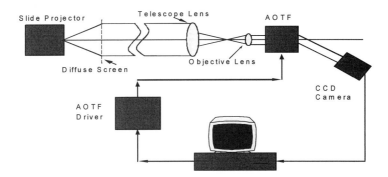

Figure 1 *Setup for using an AOTF to acquire a hyperspectral image*

3 NEURAL NETS

Commercially available neural nets were investigated and found to fall far short in terms of adequacy for this application on user friendliness and openness. Consequently, two neural nets were designed in-house, each with a "white box" approach that enabled easy access to the weights. It was reasoned that if certain features were already known to be relevant then it should be easy for the operator to insert this important information into the weight matrix. Following training, the assigned weights could be evaluated for obvious errors and any unexpected relationship could be independently verified. Using this approach, it was possible to build large robust neural nets that trained very rapidly. An unsupervised (USNN) and a supervised neural net (SNN) were devised. The USNN feeds the SNN data as shown in Figure 2.

3.1 The Unsupervised Neural Net

The USNN determines an outcome from the input data and employs a single-layer neural net. One neuron is allocated for each input wavelength in the spectrum. The USNN, following rules established by the operator, acts as a spectral sorter and sifter and is trained to determine and remember spectral characteristics.[5]

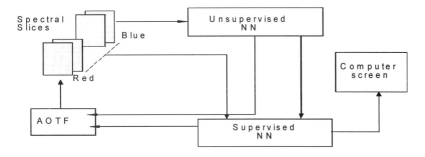

Figure 2 *System configuration*

Applying Kohonen's self-organizing feature map and its modification, the USNN operates by adjusting the interconnection weights between input and output neurons by matching a "score" between an input neuron and the memory.[6] The single-layer NN consists of NxN input and MxM output neurons forming an input and output vector space. The output neurons are extensively laterally interconnected and are also connected to the input neurons by means of weighted interconnections. In order to determine the best matching output neuron, the USNN applies similarity matching criteria to find the best match between input and memory vectors. The final output can be considered the essence of the various spectra submitted, and can be referred to as the *key spectral components* (KSC).

Each pixel acquires a complete spectrum; however, an almost unlimited number of unique spectral characteristics, such as subtle variations in intensity and wavelength ratios, may be present in a single object. The trained USNN extracts the spectrum from each pixel and determines each KSC that may be present. Spectra can be combined on this basis alone, but the operator has the option of manually combining or rejecting any spectrum that fails to be object-specific. The USNN groups the acquired spectra that are indicative of particular objects, following operator-assigned rules. In the next step, each unified spectral group is compared against all others and only those KSCs that make each group unique are selected and sent to the SNN. The process can be simplified--for example, it might be determined that there are seven objects, each with a single unique wavelength. In this case, the USNN would drive the AOTF to only generate those seven wavelengths in any future acquisition. The SNN would then be able to identify each object just on the basis of those wavelengths.

If the original image is acquired with 200 wavelengths (200 frames) and the USNN finds that the objects present can only be identified using multiple wavelengths, then the data is sent to the SNN to identify the spectral special features that are characteristic of each object.

3. 2 The Supervised Neural Net.

If a solution depends upon a large number of variables, an SNN (Table 1) can be trained to determine the relationship among those variables in the form of a weight matrix. The SNN employs multiple layers for rapid, parallel, non-linear data analysis.[7] The SNN derives the weight relationships among all variables from multiple sets of well characterized data whose outcome has been previously determined or is already known.

The "white box" structure of the SNN enables the user to input an initial set of weights known to be relevant and, if appropriate following training, permits the user to change or adjust the new weight matrix. This avoids the weakness of "black box" neural net (NN) that sometimes collapse because irrelevant relationships are established. In the worst case, except in the most simple of experiments, a black box NN may never find a local, let alone a global, minimum.

The number of variables the SNN can handle is almost unlimited, but it has been found that if training is to be rapid and effective then pre-processing the data before submitting it to the SNN is almost essential. Pre-processing algorithms may include, but not be limited to, smoothing, wavelet and Fourier transform, deconvolution, edge detection, and chemometrics. These algorithms in conjunction with the SNN form the *hybrid neural net*, which in our experience can be very robust. During training, the SNN establishes the weights as a function of special features or characteristics that have been determined to be

Table 1 *Elements of the SNN*

* Data pre-processing
* Special feature detection
* Image enhancement
* Object detection
* Labelling
* An AOTF

unique. The training method that establishes weights based on special features was developed in-house and is referred to as *inter-pattern association* (IPA). When combined with Kohonen's unsupervised feature map method of training, which establishes weights as a function of clustering of input features, the SNN is well adapted to find extremely low contrast objects. In each case, backpropagation is used to enable either process to establish weights.

3.2.1. Special Feature Detection. The SNN is fed data sets that are known to characterize specific objects. The SNN determines relationships and correlations among the wavelengths present in order to further reduce the number of frames necessary to identify an object and to speed data processing. For example, if it is determined that a given ratio between the wavelength found in frame 12 and the wavelength found in frame 59 characterizes the object, then all other wavelengths and frame acquisitions can be eliminated. Decisions concerning wavelength selection are finalized as a function of contrast enhancement.

4 SYSTEM HARDWARE

The principle of an AOTF is based on the formation of an RF-generated acoustic wave through a crystal, in this case TeO_2, that effectively creates a diffraction grating. Tuning the frequency enables a selected wavelength to be transmitted with maximum efficiency. For a hyperspectral imaging application, the area of the crystal must be large enough to permit imaging onto an array detector such as a CCD. The AOTF used in these experiments is characterized in Table 2. The CCD (Table 3), framegrabber (Table 4) and AOTF driver (Table 5) are commercial off-the-shelf hardware.

Table 2 *AOTF Characteristics*

Material of crystal	TeO_2
Aperture	8 x 10 mm
Time to fill aperture	15 µs
Response time of RF driver	15 ms
Power required	<1 W
Diffraction efficiency	>80% in polarized light.
Spectral resolution at 633 nm	3.3 nm
Spectral range	440 to 790 nm
Angular aperture	6° x 6°

Table 3 *CCD*

CCD:	Cohu model 4815
Pixels:	754 x 488
Frame rate:	30Hz

Table 4 *Framegrabber*

Framegrabber:	Matrox, 8 bit
Frame rate:	2 fr/sec

Table 5 *AOTF Driver*

AOTF driver:	Brimrose
Response time:	15 ms

5 EXPERIMENTAL RESULTS

The setup shown in Figure 1 was used to project onto a screen a slide depicting a red rose among some white flowers. A telescope captured the image of the slide, which was collected by a collimating lens, transmitted through the AOTF, and re-imaged onto the CCD. Because of the angular separation of the diffracted beams, no polarizers were required. The AOTF was driven to filter 16 wavelengths for a total data acquisition time of 0.5 sec.

The USNN was initially set up with 16 input and 16 output neurons with the weight matrix randomly initialized. The learning process depends on user-specified thresholds based on similarity measurements (i.e. Euclidean distance) between 0 and 100% so that the USNN either extracts rough or detailed spectral features. The weight matrix is shown on the computer screen and is organized into weight vectors that correspond to the output vectors. For example, the weight vector of (1,1) is the weighted connection between all the input neurons and the output neuron at the (1,1) position.

After the USNN has learned a new spectral feature, it will store the feature in one of the weight vectors. If it finds more pixels with similar spectral features (later to become a KSC), the weight vector is enhanced just as a human memory can be enhanced by repeated learning. The number of times a particular KSC is repeated in each element of the weight matrix is recorded and printed to the screen. If a particular KSC appears infrequently then this feature can be over-ridden by other features. This process is analogous to the phenomenon of human memory that we tend to forget those things that do not occur often.

If the USNN overflows, that is if every weight vector has been occupied by a spectral feature, then the USNN grows another row of weight vectors. By the time the entire HSI has been evaluated, the USNN will have learned all the major KSCs. For example, Figure 3a. shows the original image of a rose and Figure 3b the HSI of the image. The USNN determined that there were seventeen KSCs, with three neurons unoccupied. These KSCs could then either be hand selected by the user based on spectral similarity or sent in total to train the SNN. In this example, we determined that only five of the seventeen KSCs were required to identify the rose.

Figure 3a *Original input image submitted to AOTF*

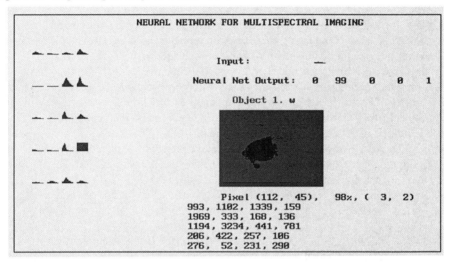

Figure 3b *Result of USNN and SNN analysis: Five key spectral components identified and represented as pseudo-color rendering.*

The SNN had sixteen input neurons, five hidden neurons and five output neurons, and was trained by backpropagation from the KSCs established by the USNN. Following training, the SNN classified each pixel into one of the five KSCs, and the confidence of the determination was printed to the screen. Using randomly selected false colors, the results were printed to screen as shown in Figure 3b. The degree of detail is determined by the user-selected similarity thresholds used to determine the number of KSCs.

The USNN required about three minutes to train and the SNN took about one minute. The recognition speed of the SNN is about 100 pixels per second. The graphical display time is negligible.

6 CONCLUSIONS

It has been shown that images of complex objects can be characterized in terms of their spectral content. The use of an acousto-optic tunable filter for wavelength selection offers the opportunity to drive the AOTF to generate combinations of wavelengths that will identify key objects of interest. Key spectral components are determined by an unsupervised neural net followed by a hybrid supervised neural net for special feature identification and mapping. Using these techniques, complex images can be rapidly evaluated using minimal memory.

ACKNOWLEDGEMENTS

Thanks to David Mintzer for his ever present help and assisitance. This work was sponsored in part under Deptartment of Navy contract number N00244-95-C-0401.

References

1. R. Anderson, W. Malila, R. Maxwell, and L. Reed, 'Military utility of multispectral and hyperspectral sensors', report of the Infrared Information Analysis Center, Environmental Research Institute of the University of Michigan, Ann Arbor, Michigan, 1994.

2. P. T. Treado, I. W. Levin, and E. N. Lewis, *App. Spec.,* 1992, **46,** 8, 1211.

3. 'Imaging Spectrometry of the Terrestrial Environment', Greg Vane (ed.), Proceedings of the Society of Photo-Optical Instrumentation Engineers, 1993, **1937**.

4. E. N. Lewis, V. F Kalasinsky, and I. W. Levin, *Anal. Chem.* 1988, **60,** 2658.

5. T. Lu, F. T. S. Yu, and D. A Gregory, *Opt. Eng.,* 1990, **29**, 9, 1107.

6. K. Kohonen, 'Self-organization and Associative Memory', Springer Verlag, New York, 1984.

7. J. M. Lerner and T. Lu, *Photonics Spectra*, August, 1993, 43.

FLEXIBLE HIGH SPEED CAMERA SYSTEM

Fritz Stauffer

National Solar Observatory
Sacramento Peak
Sunspot, NM 88349

ABSTRACT

A flexible high speed camera system with commercial off the shelf parts has been implemented at the National Solar Observatory. The current system is a Thomson 1k x 1k evaluation camera, a UNIX workstation, an RS422 digital camera interface, timer card for shutter control, and a UniBlitz shutter. The system can be run with shell scripts, from within data reduction software, or controlled via ethernet in conjunction with the telescope instruments. The Thomson camera has 10 bit data and a frame rate greater than 5 frames/second. The RS422 camera interface allows for 8 to 16 bit camera data with variable image formats at greater than 10 million pixels per second. An image can be selected according to contrast, binned, windowed and averaged before storage. For a 1k x 1k image, frame averaging is greater than 4 frames per second, and storage rates are 2 frames per second. The timer card allows the system to be externally synched to an experiment, and controls the shutter timing. The system versatility allows other digital cameras with an RS422 interface to easily be integrated.

Subject Index